AF679452

Structural Analysis II
Homework and Exams

Peter I. Kattan

Petra Books
www.PetraBooks.com

Peter I. Kattan, PhD

Correspondence about this book may be sent to the author at one of the following two email addresses:

pkattan@petrabooks.com

info@petrabooks.com

Structural Analysis II. Homework and Exams
ISBN: 979-8-3493-0230-5
All rights reserved. No part of this book may be copied or reproduced without written permission of the author or publisher.
© 2025 Peter I. Kattan

Preface

These are the homework sets and exam sets for the course Structural Analysis II that was taught by Dr. Peter I. Kattan. These homework and exam sets are a follow-up volume to the successful volume Structural Analysis II Lecture Notes that is available online and at fine bookstores.

This volume includes five homework sets and seven exam sets. Solutions in detail are provided for some of these sets. In particular, detailed handwritten solutions are provided for one homework set and for two exam sets. Note that these sets as well as the associated Lecture Notes are for the second course in structural analysis that covers indeterminate structures. Another set of notes as well as homework and exams are available for the first course on determinate structures.

The author makes these notes and sets available as free downloads as PDF files on the internet. However, the printed paperback copies are provided at a greatly reduced price at fine bookstores online and everywhere. The total size of this volume is about 50 pages. Not all the solutions are provided but enough is given to keep the student busy for hours.

April 2025 Peter I. Kattan

Structural Analysis II
Homework and Exams

Contents

Homework 1	7
Homework 3	8
Homework 4	9
Homework 6	10
Homework 7	11
Exam 1	12
Exam 1	13
Exam 1	15
Exam 2	16
Exam 2	17
Exam 2	19
Final Exam	20
Solutions to Homework 7	22
Solutions to Exam 2	30
Solutions to Final Exam	33

Structural Analysis II
Homework and Exams

APPLIED SCIENCE UNIVERSITY
FACULTY OF ENGINEERING

Structural Analysis (2)

Homework 1

1- Problem 11.16 page 249 in the textbook (Alexander Chajes).

2- Problem 11.20 page 250 in the textbook (Alexander Chajes).

3- Problem 11.26 page 251 in the textbook (Alexander Chajes).

4- Problem 11.30 page 252 in the textbook (Alexander Chajes).

5- Determine the member forces in the truss shown in the figure below. The cross-sectional area of all top chord members is 2A and the cross-sectional area of the remaining members is A. Take E = constant. The diagonal members have no joints at the points where they cross.

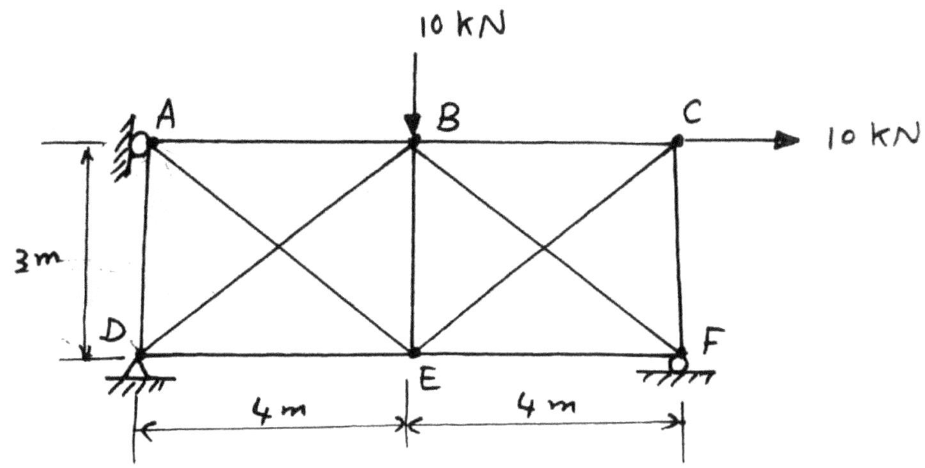

Answer to Problem 5:

F_{AB} = 6.21 kN (T), F_{BC} = 6.28 kN (T), F_{DE} = 3.06 kN (T), F_{EF} = 2.99 kN (T),
F_{AD} = 2.75 kN (C), F_{BE} = 5.55 kN (C), F_{CF} = 2.80 kN (C), F_{AE} = 4.58 kN (T),
F_{BD} = 3.66 kN (C), F_{CE} = 4.67 kN (T), F_{BF} = 3.75 kN (C)

APPLIED SCIENCE UNIVERSITY
FACULTY OF ENGINEERING

Structural Analysis (2)

Homework 3

1. For the continuous beam shown in the figure, determine the reactions and draw the beding moment diagram. EI = constant.

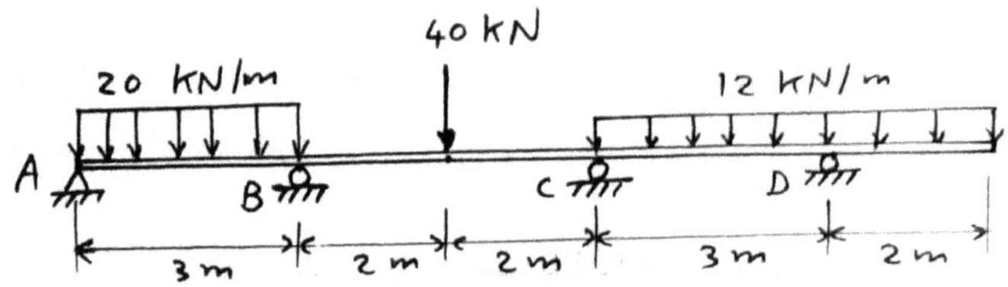

2. The continuous beam shown in the figure has a constant I value of 250 x 10^6 mm^4 and E = 200 GPa. Determine the reactions and draw the bending moment diagram if the support at B settles 8 mm.

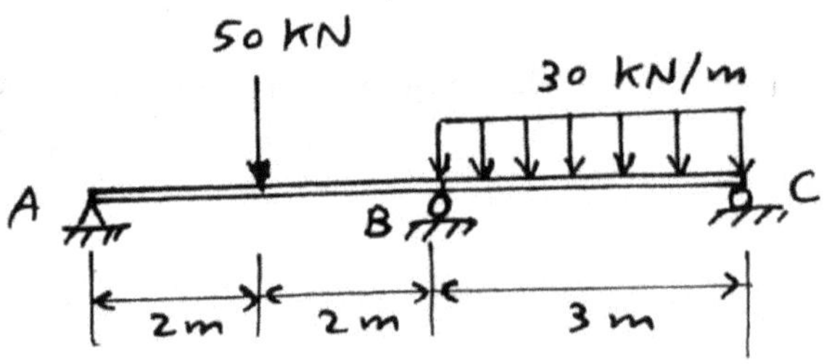

3. Determine the support moments in the continuous beam shown in the figure.

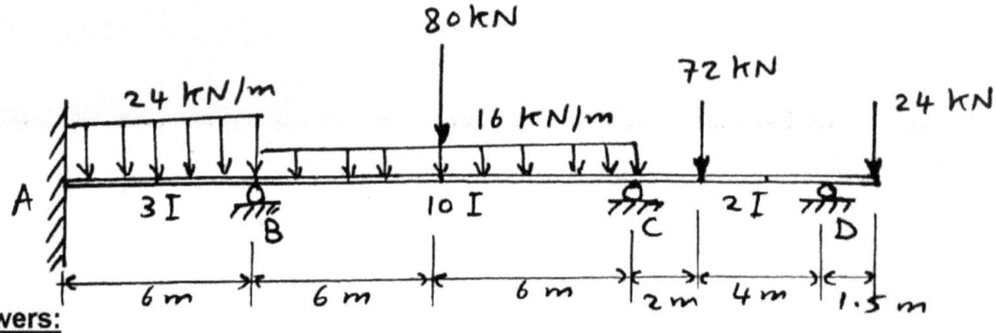

Answers:

1. R_A = 22.12 kN, R_B = 61.03 kN, R_C = 30.53 kN, R_D = 46.32 kN.
2. R_A = 41.03 kN, R_B = 32.60 kN, R_C = 66.37 kN.
3. M_A = -0.36 kN.m, M_B = -215.28 kN.m, M_C = -147.25 kN.m

APPLIED SCIENCE UNIVERSITY
FACULTY OF ENGINEERING

Structural Analysis (2)

Homework 4

1- Problem 13.4 page 298 in the textbook (Alexander Chajes).

2- Problem 13.8 page 299 in the textbook (Alexander Chajes).

3- Problem 13.12 page 300 in the textbook (Alexander Chajes).

4- Problem 13.20 page 302 in the textbook (Alexander Chajes).

5- Determine the moments at each joint of the frame shown in the figure.
EI is constant for each member.

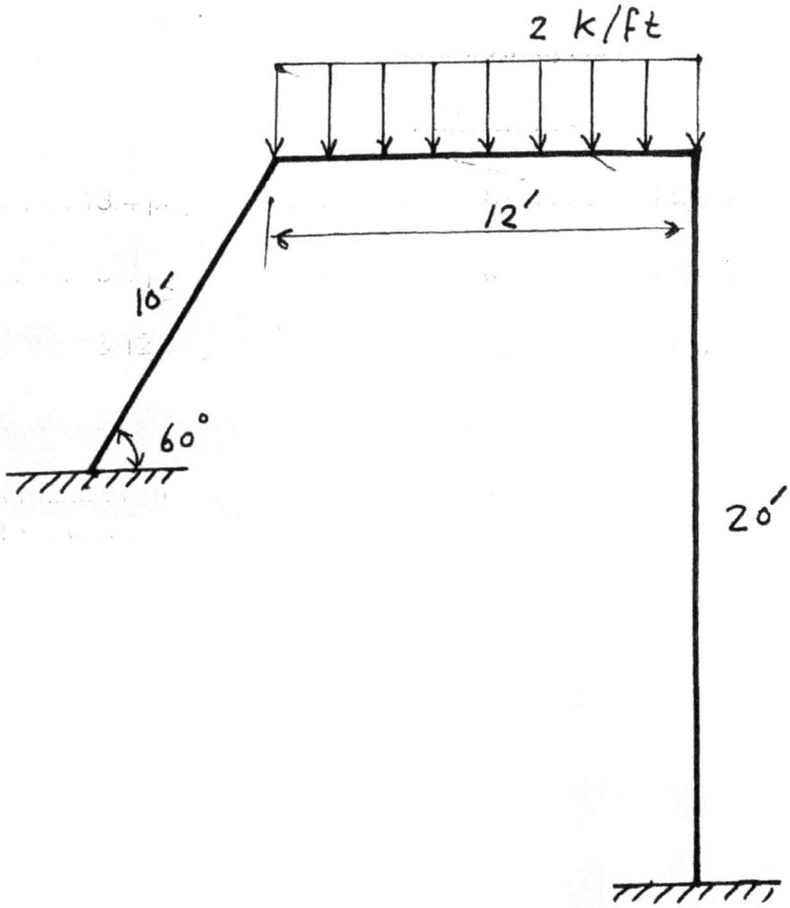

APPLIED SCIENCE UNIVERSITY
FACULTY OF ENGINEERING

Structural Analysis (2)

Homework 6

1- Determine the moments at each joint of the frame shown in the figure. Take EI = constant.

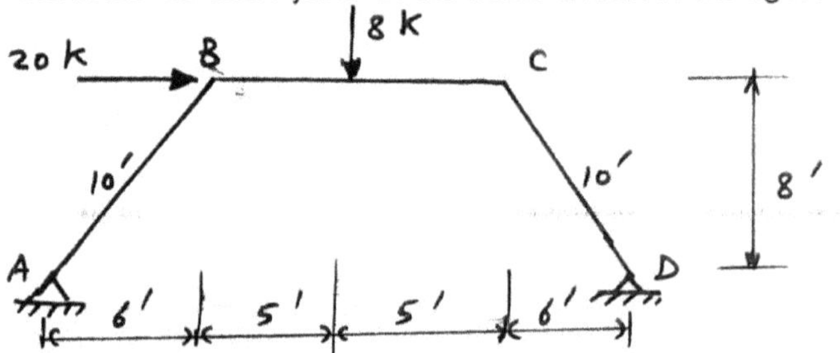

2- Determine the fixed-end moments of the beam shown in the figure below. The cross-sectional area has a constant width of 1 ft. Take E = constant.

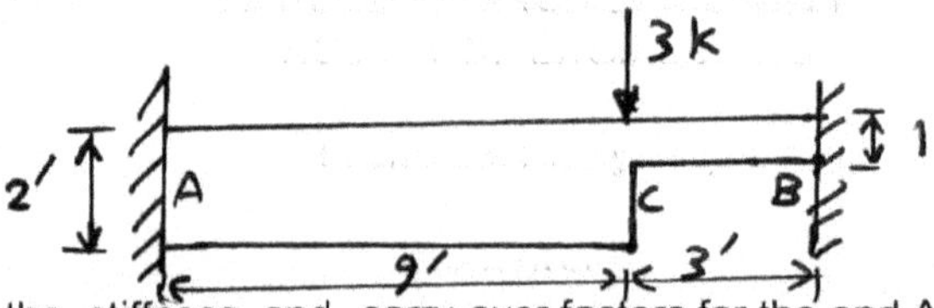

3- Determine the stiffness and carry-over factors for the end A of the beam shown in the figure below. The cross-sectional area has a constant width of 1 ft.
Take E = 10 x 10^3 ksi.

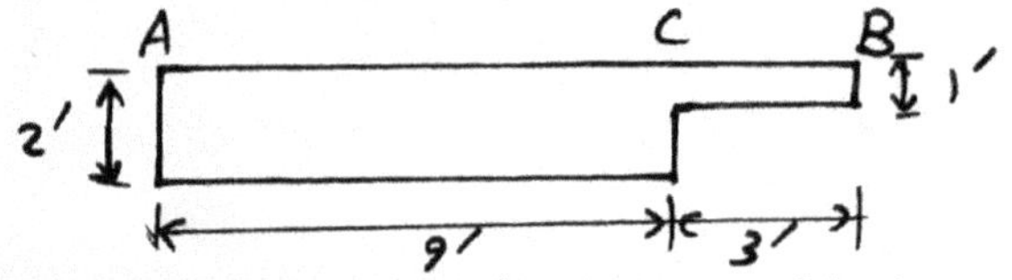

4- Determine the internal moments at the supports of the beam shown in the figure below. The beam has parabolic haunches and a thickness of 1 ft. and E is constant.

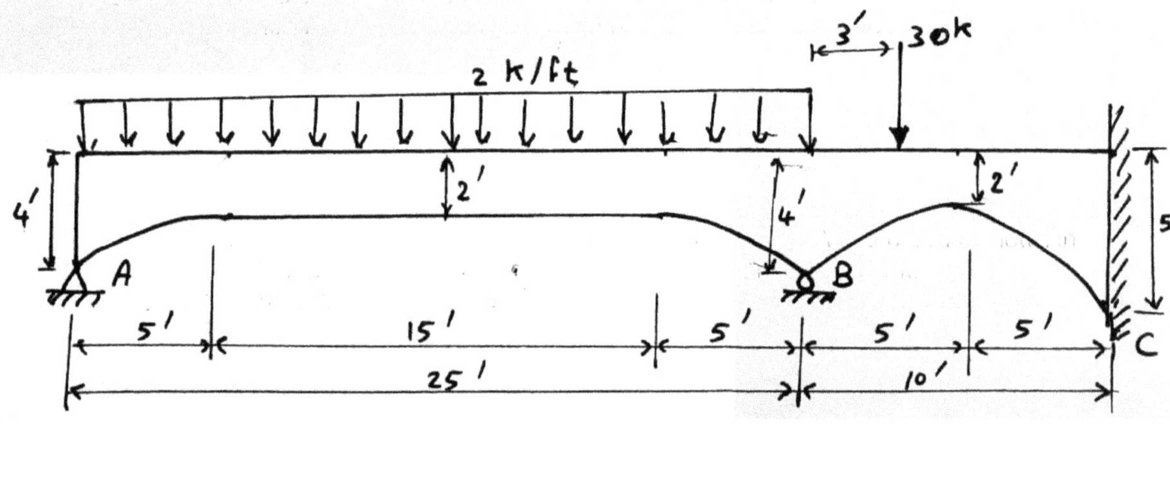

APPLIED SCIENCE UNIVERSITY
FACULTY OF ENGINEERING

Structural Analysis (2)

Homework 7

1- Analyze the box frame shown in the figure by the moment distribution method. Draw the bending moment diagram. Assume EI = constant.

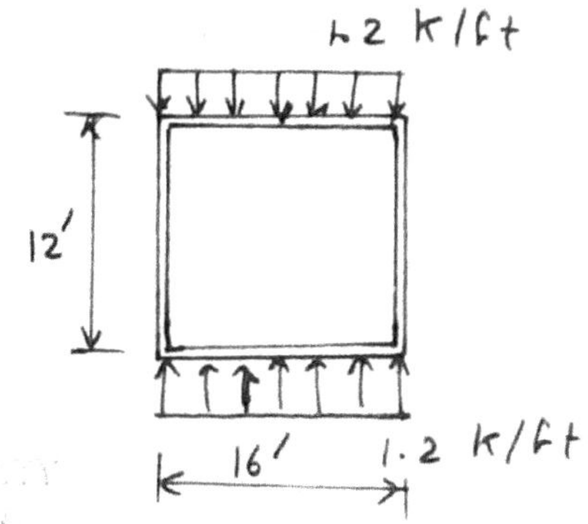

2- Consider the continuous beam shown in the figure.
(a) Compute the ordinates at 2-ft intervals of the influence line for the moment at point D. Draw the influence line to scale.
(b) Compute the ordinates at 2-ft intervals of the influence line for the shear at point D. Draw the influence line to scale.

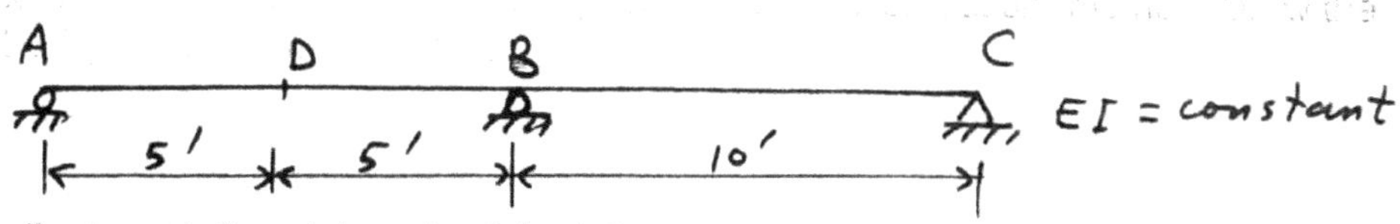

3- Compute the ordinates at 2-m intervals of the influence line for the reaction at A of the beam shown in the figure. Draw the influence line to scale.

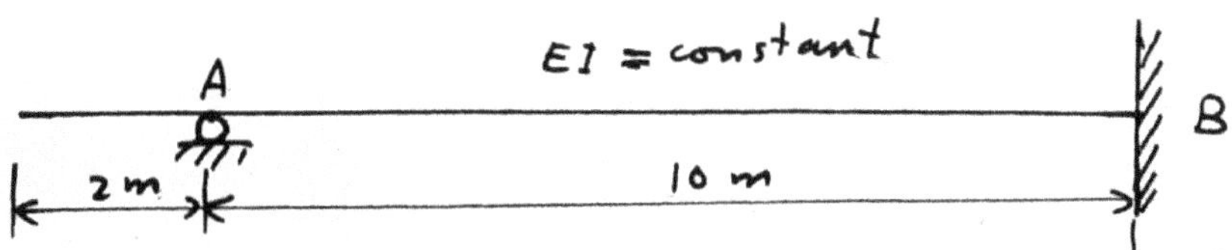

4- Consider the continuous beam shown in the figure. Draw qualitative influence lines for:
(a) the reactions at A and B.
(b) the shear just to the left of B.
(c) the shear just to the right of B.
(d) the moment at B.
(e) the shear and moment at a section in the middle of span DE.

Applied Science University
Faculty of Engineering

Dr. Peter I. Kattan
Time: 1 hour

Structural Analysis (2)
Exam 1

1. (50%) Draw the bending moment diagram for the beam shown in the figure. Take EI = constant. Use R_B as the redundant.

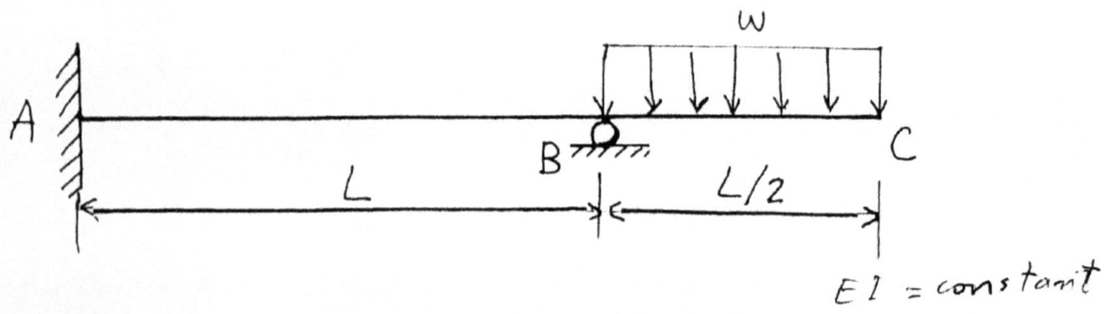

2. (50%) For the truss shown in the figure, determine all the bar forces. Use F_{BD} as the redundant.

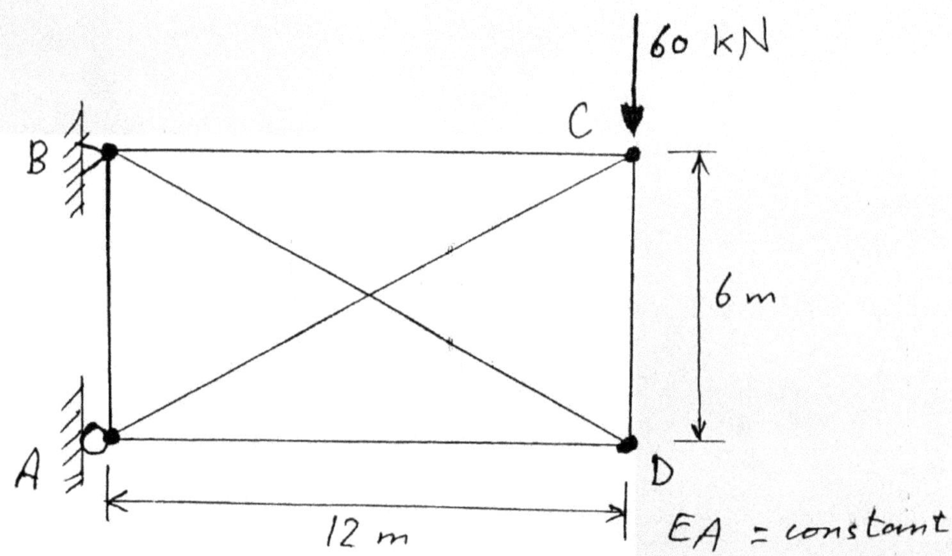

Applied Science University
Faculty of Engineering

Dr. Peter I. Kattan
Time: 1:15 hours

Structural Analysis (2)
Exam 1

1. (25%) For the structure shown in the figure, use the method of consistent deformations to find the force in the spring. Take E = 20,000 kN/cm^2, I = 5000 cm^4, k = 5 kN/cm, (EI = constant).

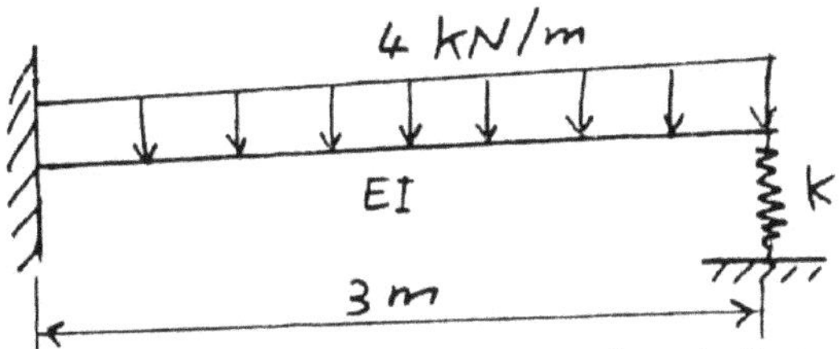

2. (25%) Use the method of least work (Castigliano's Second Theorem) to determine the reactions and draw the bending moment diagram for the semicircular two-hinged arch shown in the figure, (EI = constant).

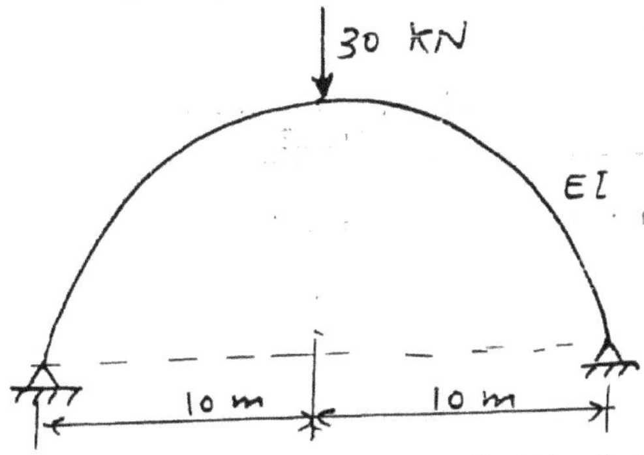

3. (15%) Use the three-moment equation to prove that the fixed-end moments of the beam shown in the figure are equal to wL^2/12, (EI = constant).

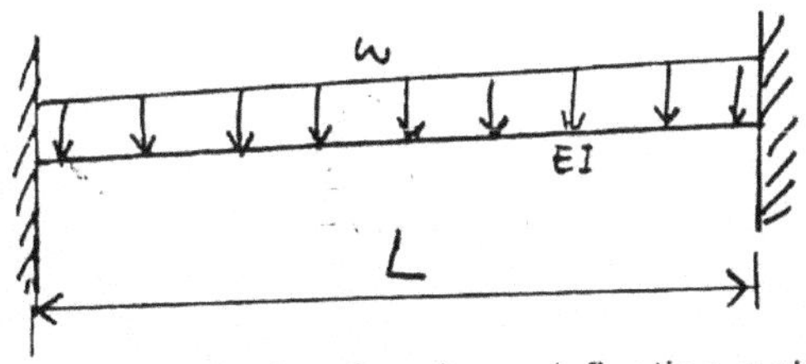

4. (25%) Solve problem 1 above using the slope-deflection method.

5. (10%) Draw the bending moment diagram for the beam shown in the figure if support B settles 1 cm. Take E = 20,000 kN/cm^2, I = 5000 cm^4, (EI = constant). Use the fastest possible method to solve the problem.

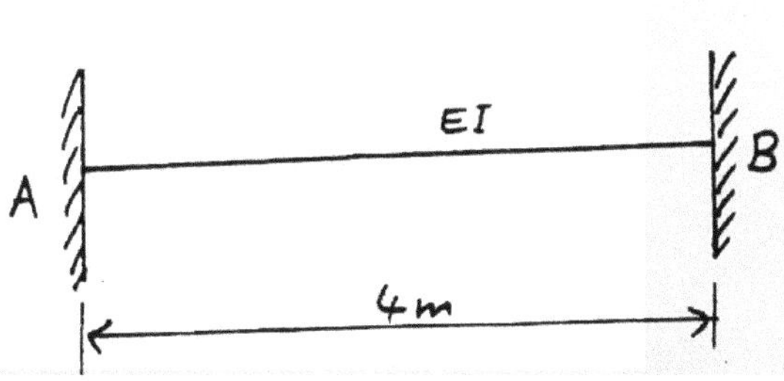

Fixed-End Moments

$\text{FEM}_{AB} = \dfrac{PL}{8}$ 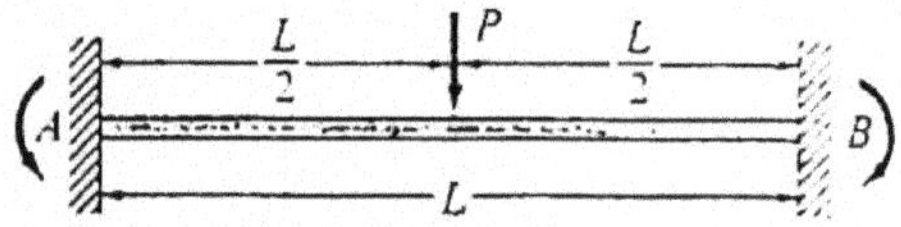 $\text{FEM}_{BA} = \dfrac{PL}{8}$

$\text{FEM}_{AB} = \dfrac{Pb^2a}{L^2}$ 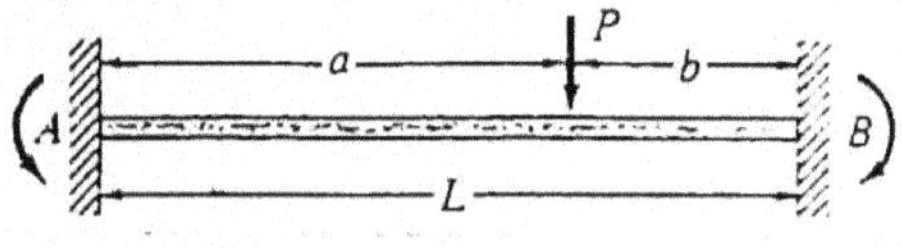 $\text{FEM}_{BA} = \dfrac{Pa^2b}{L^2}$

$\text{FEM}_{AB} = \dfrac{wL^2}{12}$ 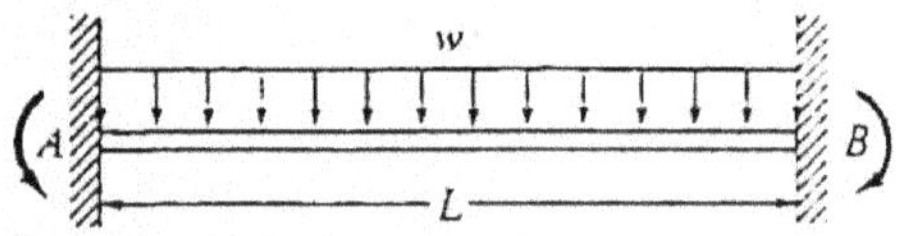 $\text{FEM}_{BA} = \dfrac{wL^2}{12}$

$\text{FEM}_{AB} = \dfrac{11wL^2}{192}$ 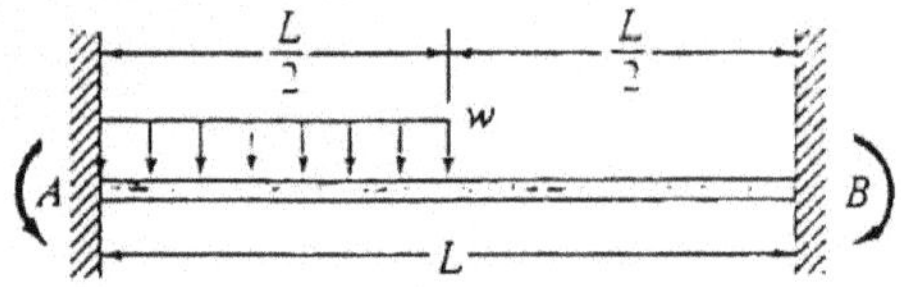 $\text{FEM}_{BA} = \dfrac{5wL^2}{192}$

$\text{FEM}_{AB} = \dfrac{wL^2}{20}$ 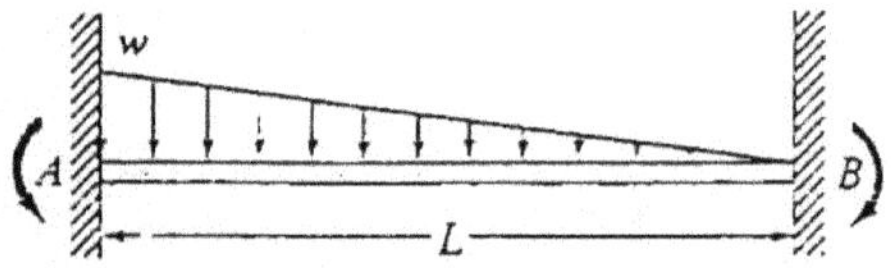 $\text{FEM}_{BA} = \dfrac{wL^2}{30}$

$\text{FEM}_{AB} = \dfrac{5wL^2}{96}$ 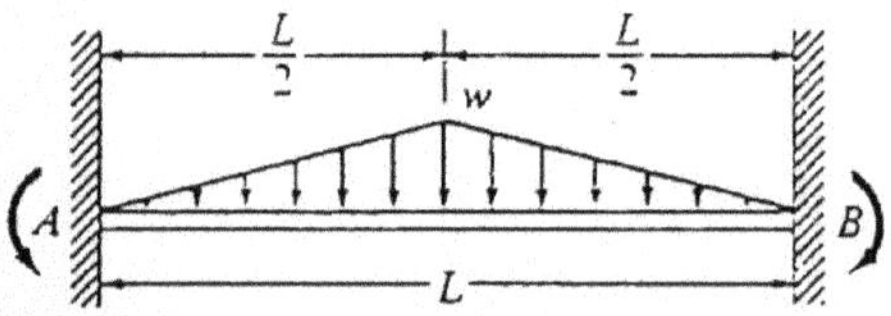 $\text{FEM}_{BA} = \dfrac{5wL^2}{96}$

$\text{FEM}_{AB} = \dfrac{6EI\Delta}{L^2}$ 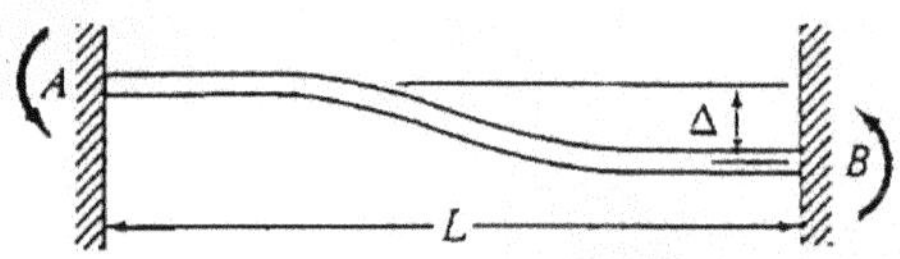 $\text{FEM}_{BA} = \dfrac{6EI\Delta}{L^2}$

$\text{FEM}_{AB} = M(1-k)(1-3k)$ 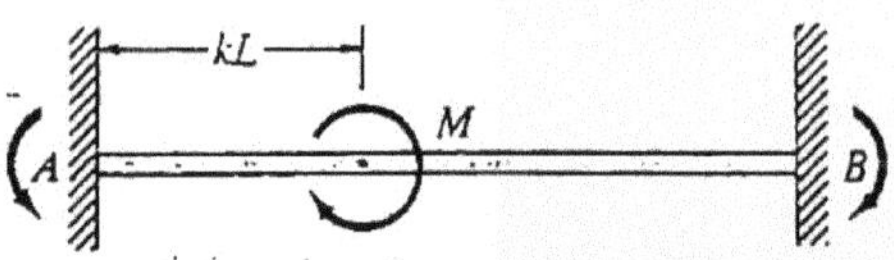 $\text{FEM}_{BA} = Mk(2-3k)$

$\text{FEM}_{AB} = \dfrac{wL^2}{60} \cdot k^2(10 - 10k + 3k^2)$ 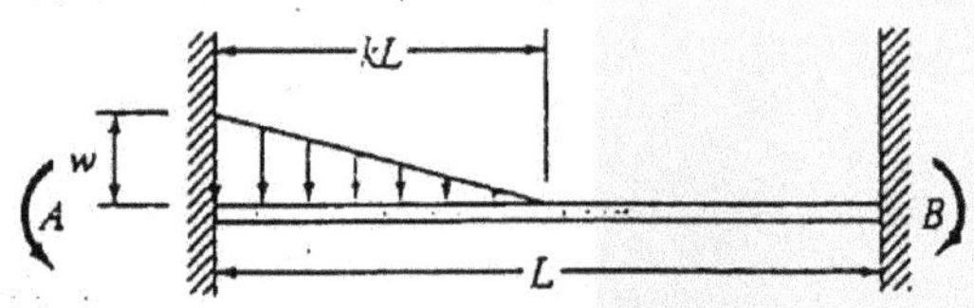 $\text{FEM}_{BA} = \dfrac{wL^2}{60} \cdot k^3(5-3k)$

Applied Science University
Faculty of Engineering

Dr. Peter I. Kattan
Time: 1 hour

Structural Analysis (2)
Exam 1

(50%) Using the *method of consistent deformations*, determine the forces in the three members of the truss shown in the figure. Take EA to be constant.

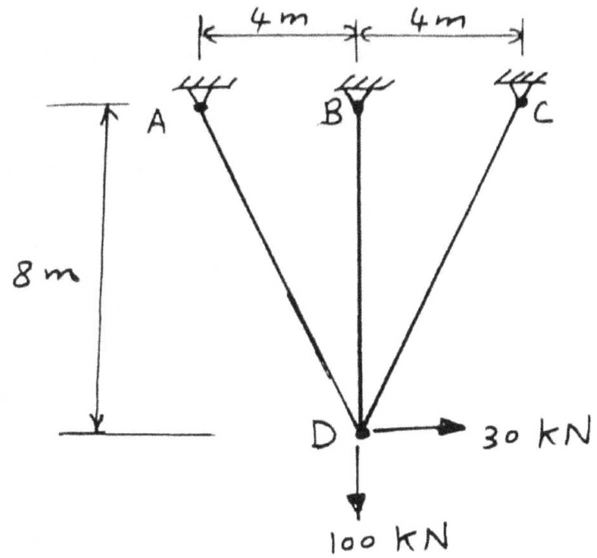

2. (50%) Using the *method of least work* (Castigliano's Second Theorem), draw the bending moment diagram for the closed ring shown in the figure.

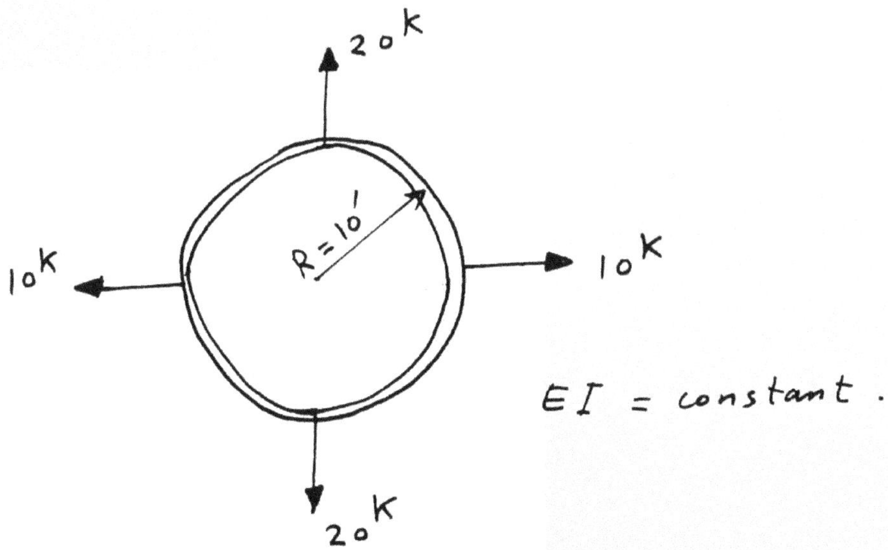

Applied Science University
Faculty of Engineering

Dr. Peter I. Kattan
Time: 1 hour

Structural Analysis (2)
Exam 2

1. (50%) Determine the force in each member of the truss shown in the figure. Use $E = 29 \times 10^3$ ksi, $A = 50$ in^2, and $I = 300$ in^4.

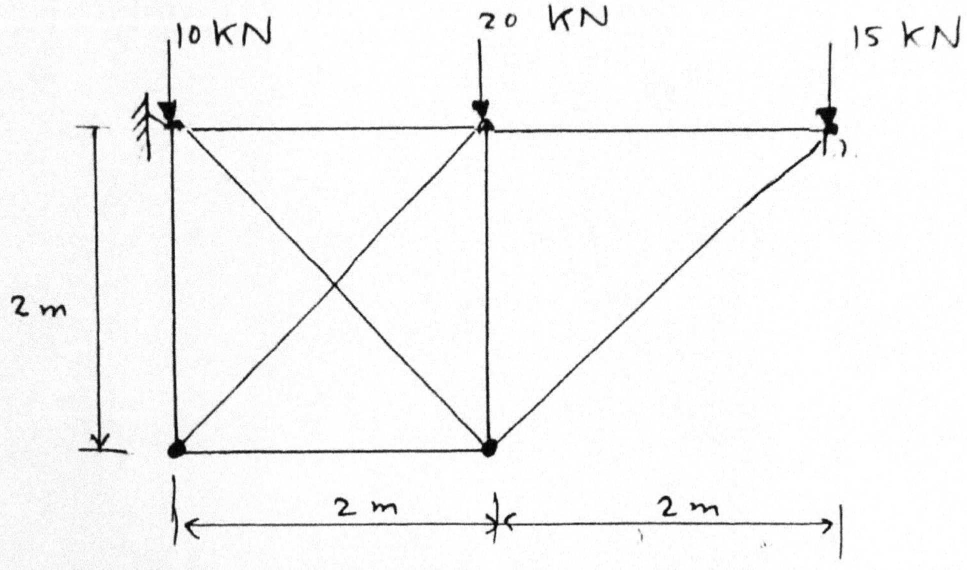

2. (50%) Determine the moments at A, B, and C for the beam with nonprismatic members shown in the figure. The beam has a thickness of 1 ft. Use $E = 29 \times 10^3$ ksi.

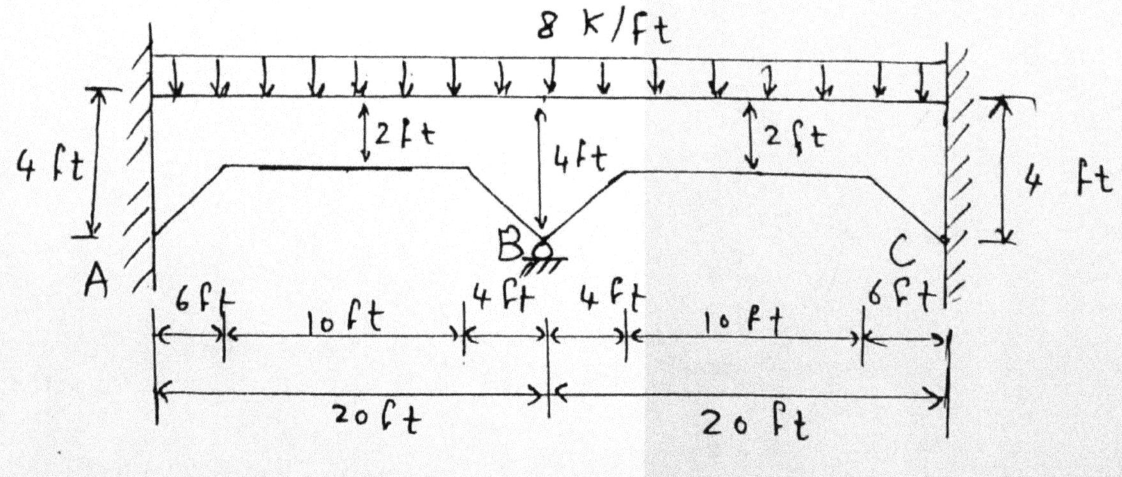

Applied Science University
Faculty of Engineering

Dr. Peter I. Kattan
Time: 1 hour

Structural Analysis (2)
Exam 2

(50%) Using the *slope deflection method*, determine the member moments for the frame shown in the figure. Draw the bending moment diagram. Assume EI = constant.

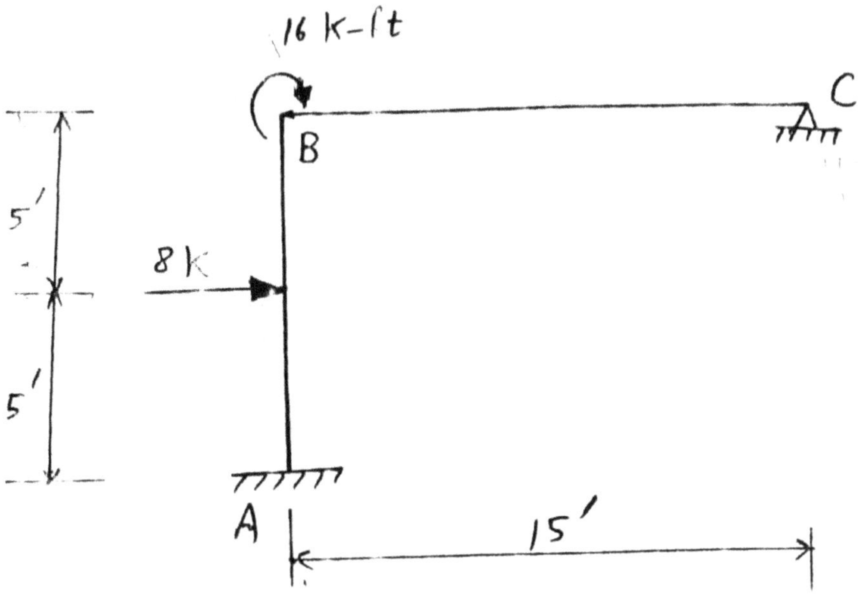

2. (50%) Using the *moment distribution method*, determine the member moments for the beam shown in the figure. Assume EI = constant.

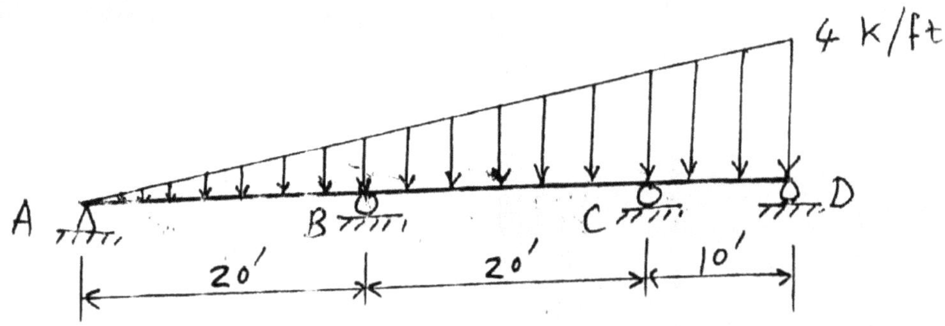

Fixed-End Moments

$FEM_{AB} = \frac{PL}{8}$

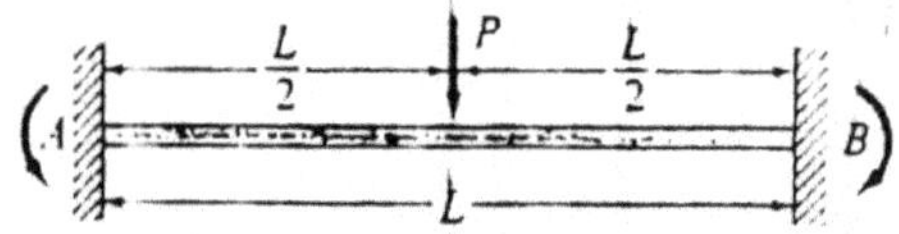

$FEM_{BA} = \frac{PL}{8}$

$FEM_{AB} = \frac{Pb^2a}{L^2}$

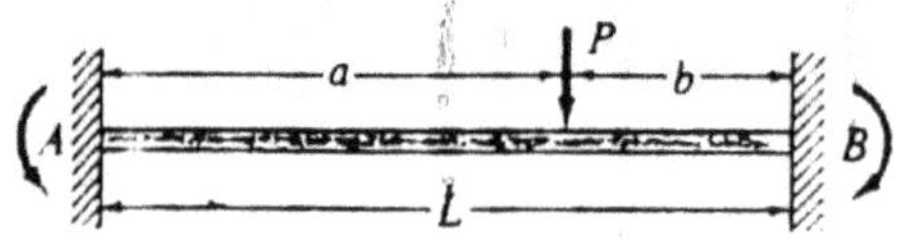

$FEM_{BA} = \frac{Pa^2b}{L^2}$

$FEM_{AB} = \frac{wL^2}{12}$

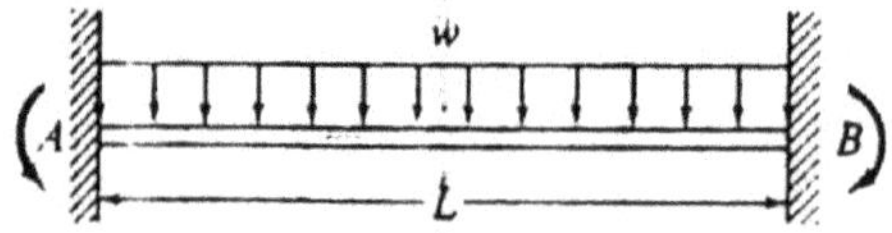

$FEM_{BA} = \frac{wL^2}{12}$

$FEM_{AB} = \frac{11wL^2}{192}$

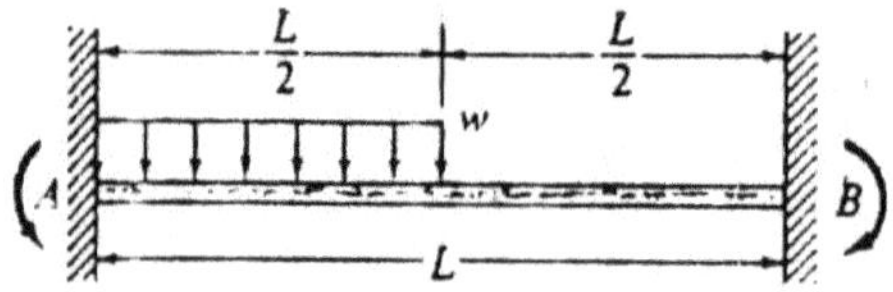

$FEM_{BA} = \frac{5wL^2}{192}$

$FEM_{AB} = \frac{wL^2}{20}$

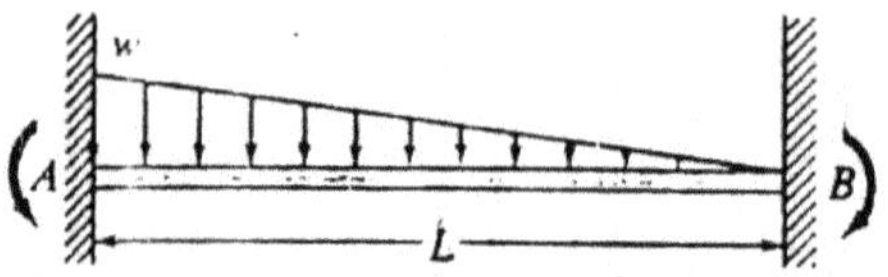

$FEM_{BA} = \frac{wL^2}{30}$

$FEM_{AB} = \frac{5wL^2}{96}$

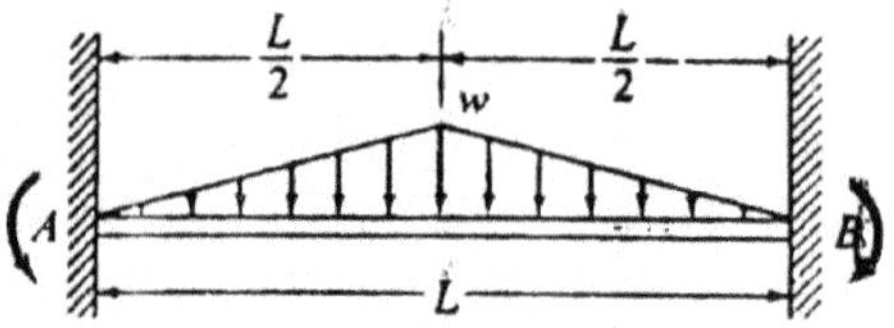

$FEM_{BA} = \frac{5wL^2}{96}$

$FEM_{AB} = \frac{6EI\Delta}{L^2}$

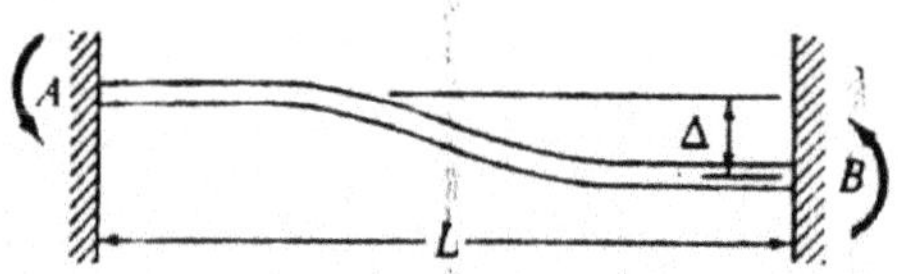

$FEM_{BA} = \frac{6EI\Delta}{L^2}$

$FEM_{AB} = M(1-k)(1-3k)$

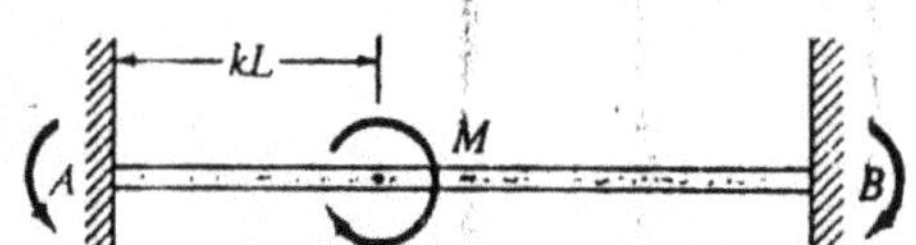

$FEM_{BA} = Mk(2-3k)$

$FEM_{AB} = \frac{wL^2}{60} \cdot k^2(10 - 10k + 3k^2)$

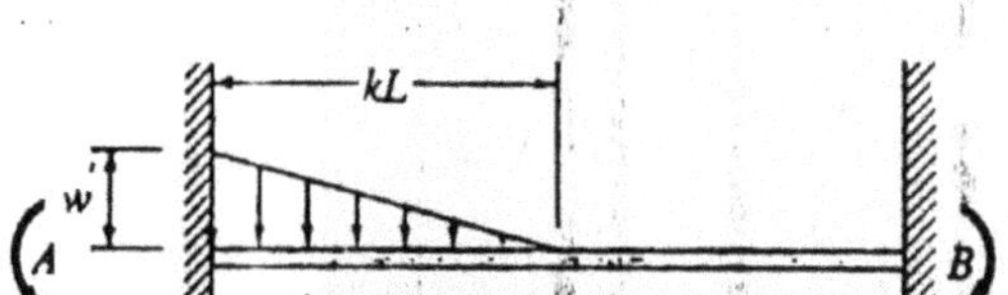

$FEM_{BA} = \frac{wL^2}{60} \cdot k^3(5-3k)$

Applied Science University
Faculty of Engineering

Dr. Peter I. Kattan
Time: 50 min.

Structural Analysis (2)
Exam 2

1. (25%) Using the slope-deflection method, write the shear equation for the frame shown in the figure in terms of the joint moments. Do ***not*** solve the problem.

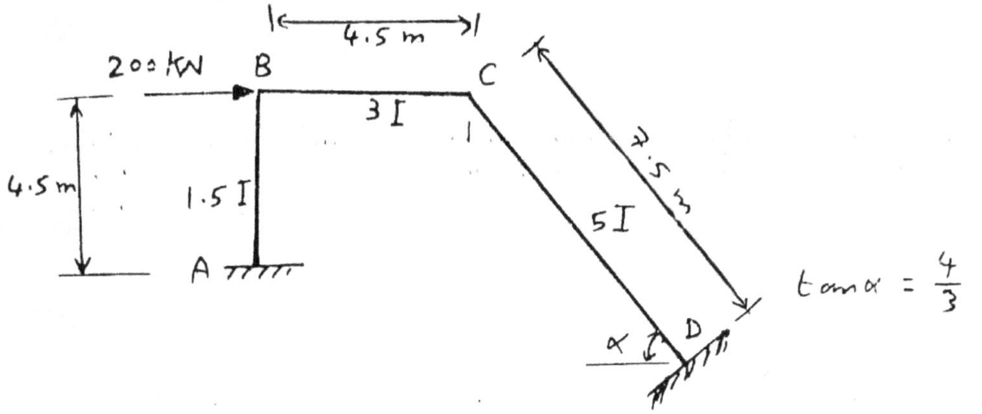

2. (35%) Draw the bending moment diagram for the frame shown in the figure.

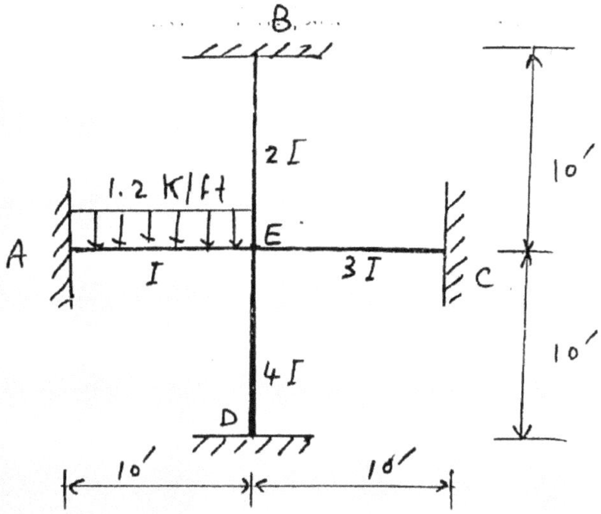

3. **(40%) Find the equation y = f(x) of the influence line for the reaction at support B of the continuous beam shown in the figure. Do *not* calculate the ordinates of the influence line or plot the curve. Take EI = constant.**

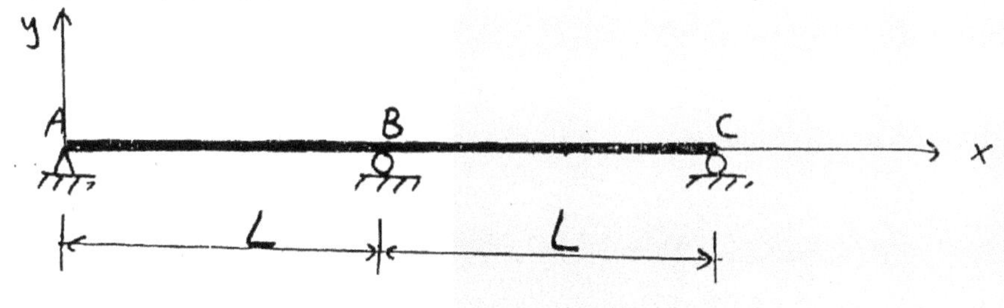

Applied Science University
Faculty of Engineering
23-6-1997

Dr. Peter I. Kattan
Time: 2 hours

Structural Analysis (2)
Final Exam

Answer all questions:

1- (25%) Determine the fixed-end moments for the beam shown in the figure. Neglect axial forces. Take EI = constant. (Hint: *Use the method of consistent deformations*).

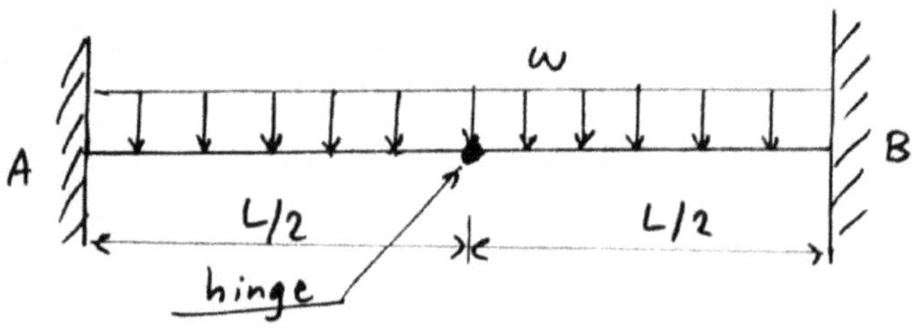

2- (20%) Draw the bending moment diagram for the two-hinged semi-circular arch with radius 15 m as shown in the figure. Take EI = constant. (Hint: *Use the method of least work - Castigliano's second theorem*).

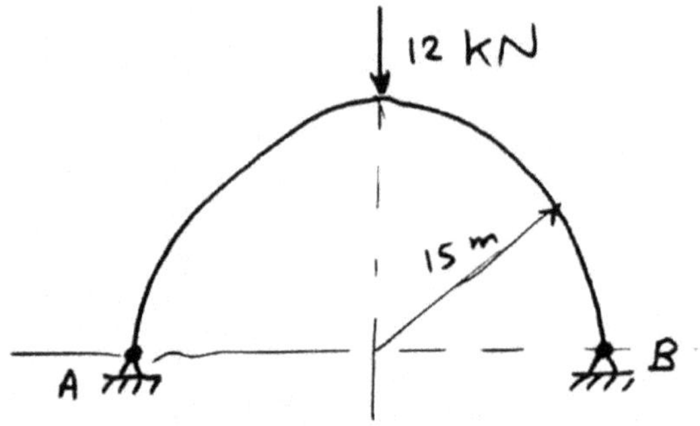

3- (15%) Using the slope-deflection method, write the shear equation in terms of the joint moments for the frame shown in the figure. Do not try to solve the problem.

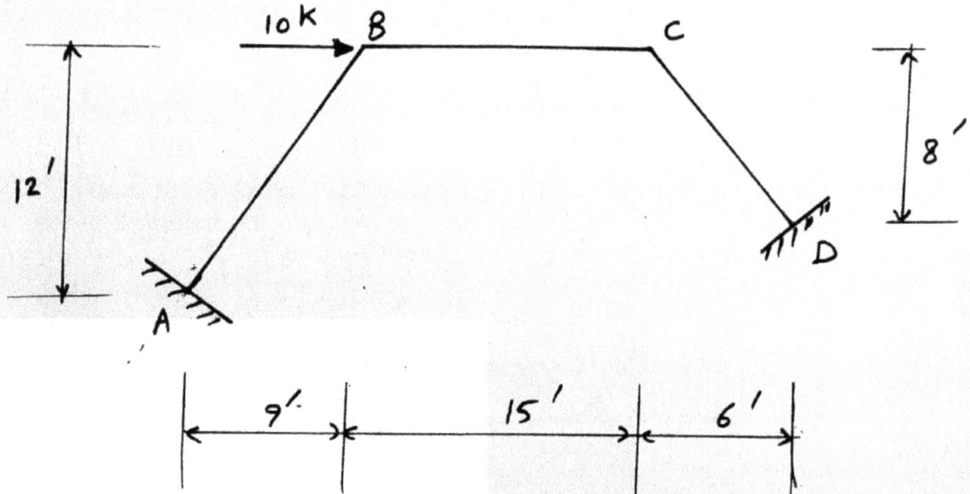

4- (15%) Draw the bending moment diagram for the frame shown in the figure. Take EI = constant. (Hint: *Use the moment distribution method*).

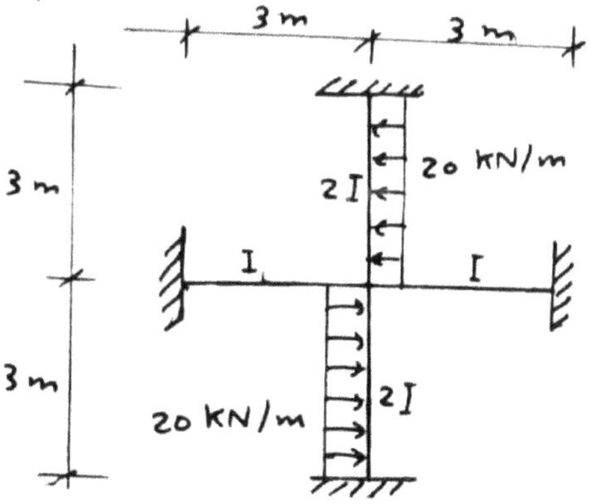

5- (25%) Draw the influence line for the reaction at A showing numerical ordinates at 1 m intervals for the continuous beam shown in the figure. Take EI = constant.

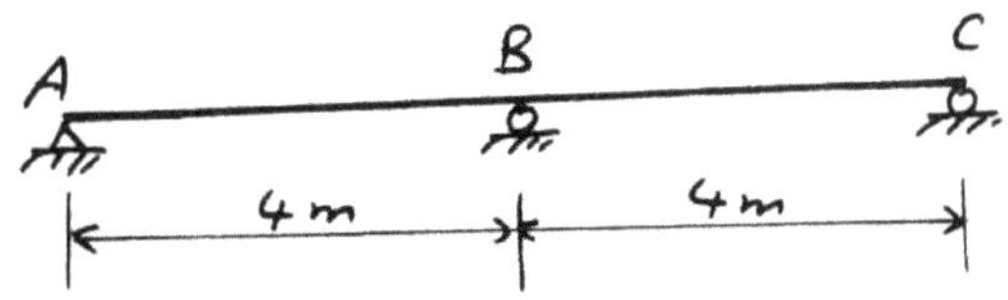

Bonus: Answer the following question for an additional 10%

For each of the following structures, determine:

(a) degree of statical indeterminacy.
(b) degree of kinematical indeterminacy.

(1)

(2)

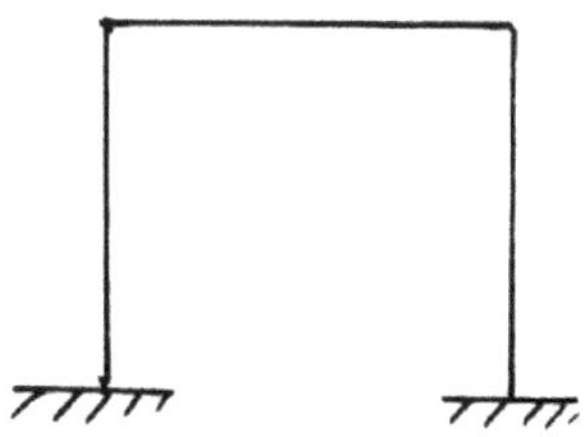

(3)

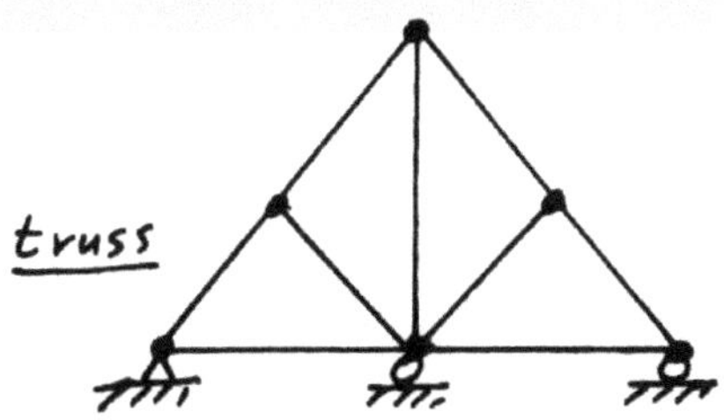

(4)

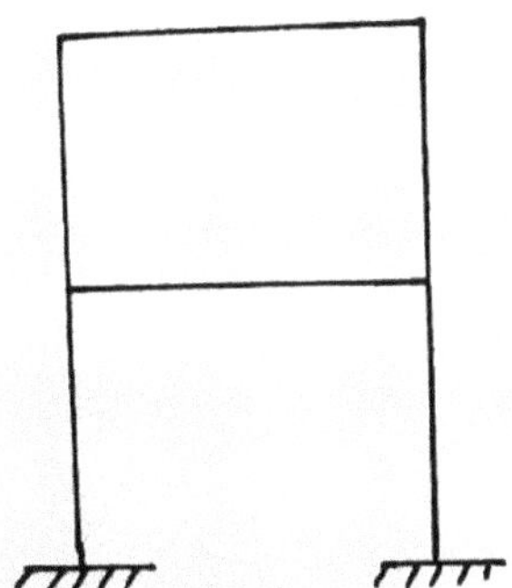

(5)

Homework 7

LH1

① Analyze the box frame shown in the figure by the moment distribution method. Draw the bending moment diagram. Assume EI = constant.

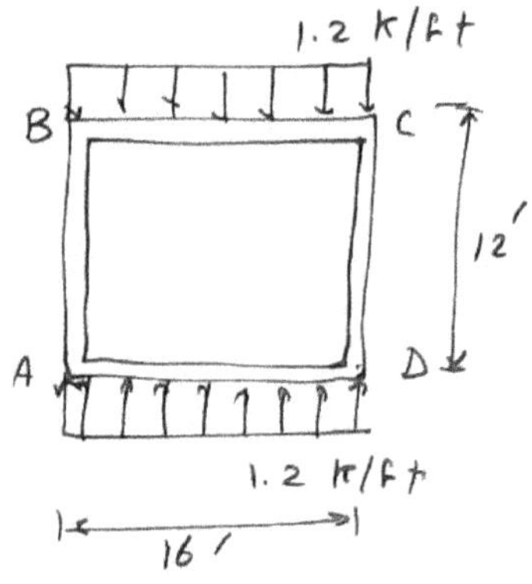

Solution:

Joint B:

$$K_{BC} = \frac{I}{16}, \quad K_{BA} = \frac{I}{12}$$

$$\Sigma K_B = \frac{I}{16} + \frac{I}{12} = \left(\frac{3+4}{48}\right) I = \frac{7I}{48}$$

$$DF_{BC} = \frac{I/16}{7I/48} = \frac{1}{16} \cdot \frac{48}{7} = \frac{3}{7}$$

$$DF_{BA} = \frac{I/12}{7I/48} = \frac{1}{12} \cdot \frac{48}{7} = \frac{4}{7}$$

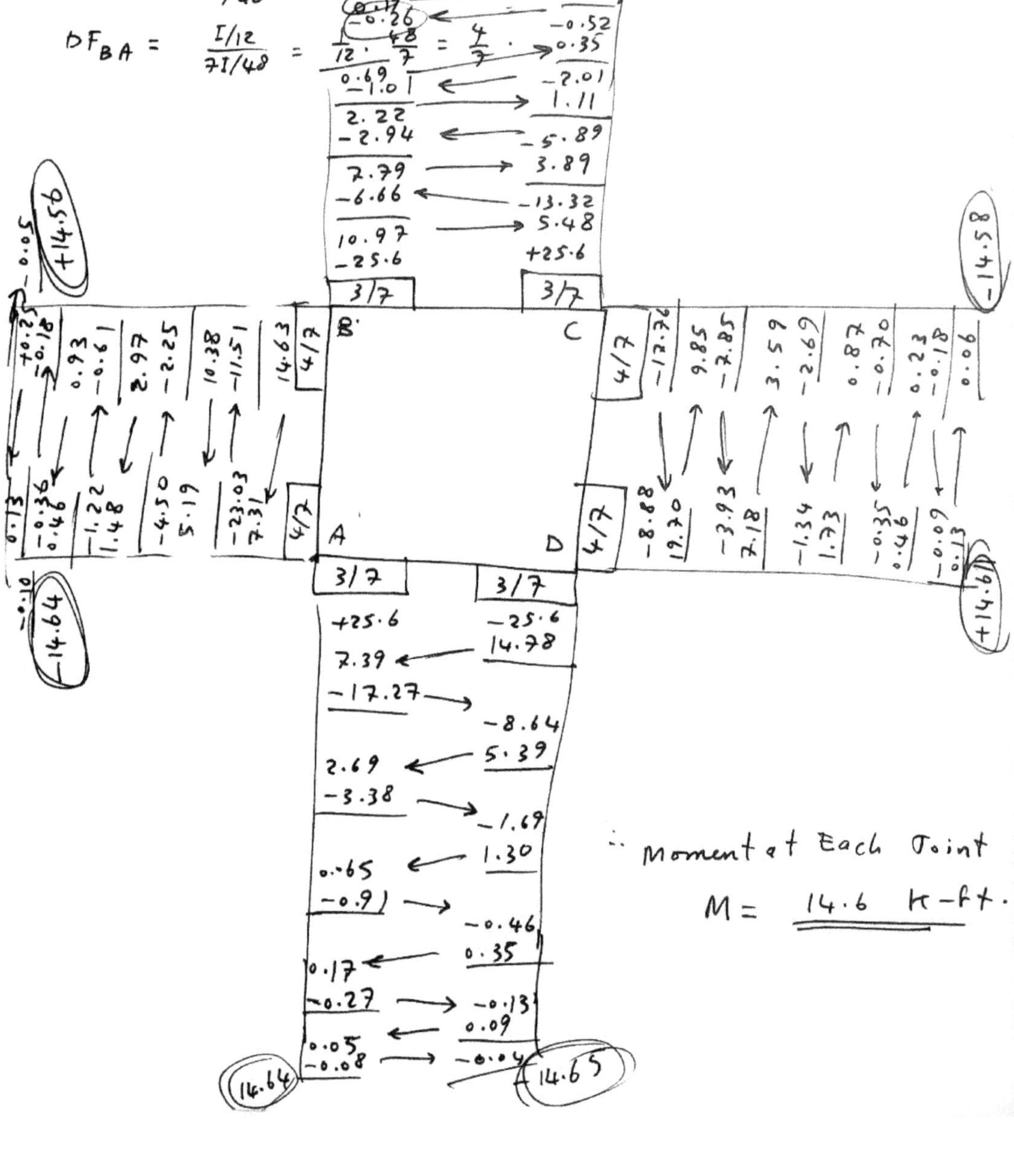

∴ Moment at Each Joint

$M = 14.6$ k-ft.

(H2

Bending Moment Diagram:

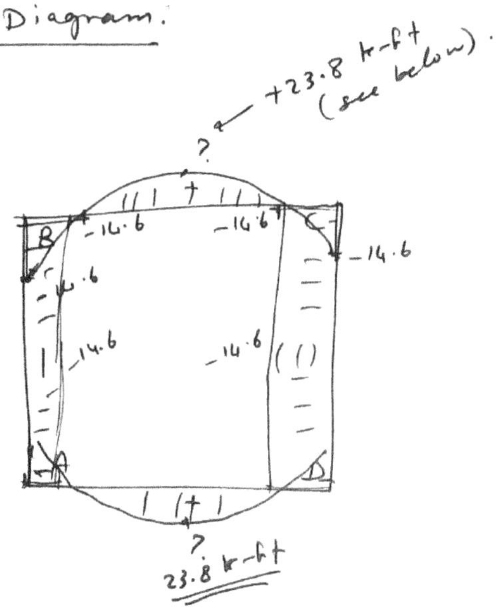

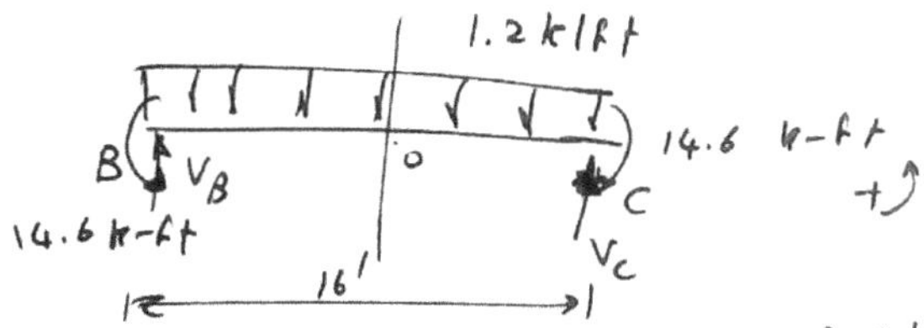

$+\circlearrowleft \Sigma M_B = 0:$

$$V_C(16) + 14.6 - 14.6 - 1.2(16)\left(\frac{16}{2}\right) = 0$$

$$V_C = 9.6 \text{ k.}$$

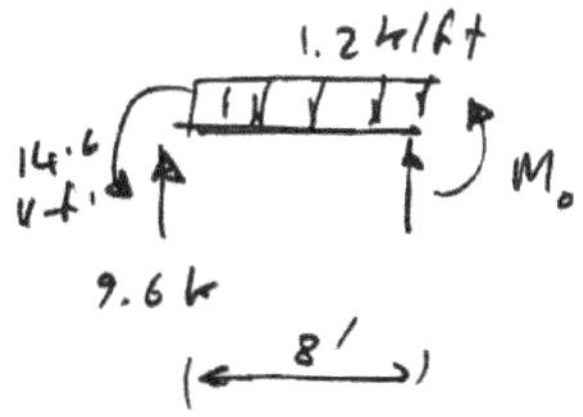

$$M_o = -14.6 - 1.2(8)\left(\frac{8}{2}\right) + 9.6(8)$$

$$= \underline{+23.8} \text{ k-ft.}$$

---

(2)

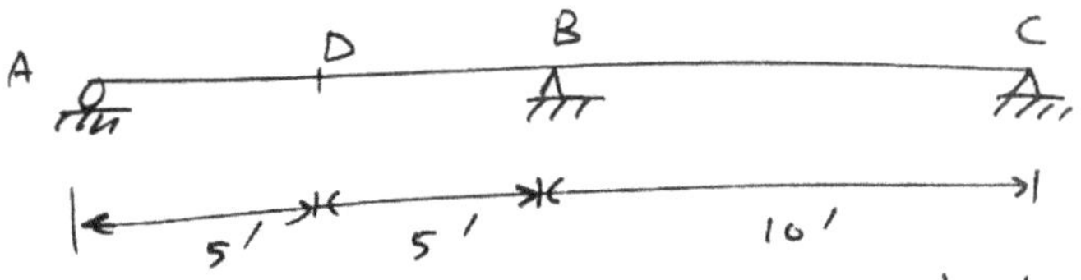

EI = constant

(a) Compute the ordinates at 2-ft intervals of the influence line for the moment at point D of the beam shown in the figure.

(b) Compute the ordinates at 2-ft intervals of the influence line for the shear at point D of the beam shown in the figure. Draw both I.L. to scale.

Solution: See attached solution.

Problem 2-a

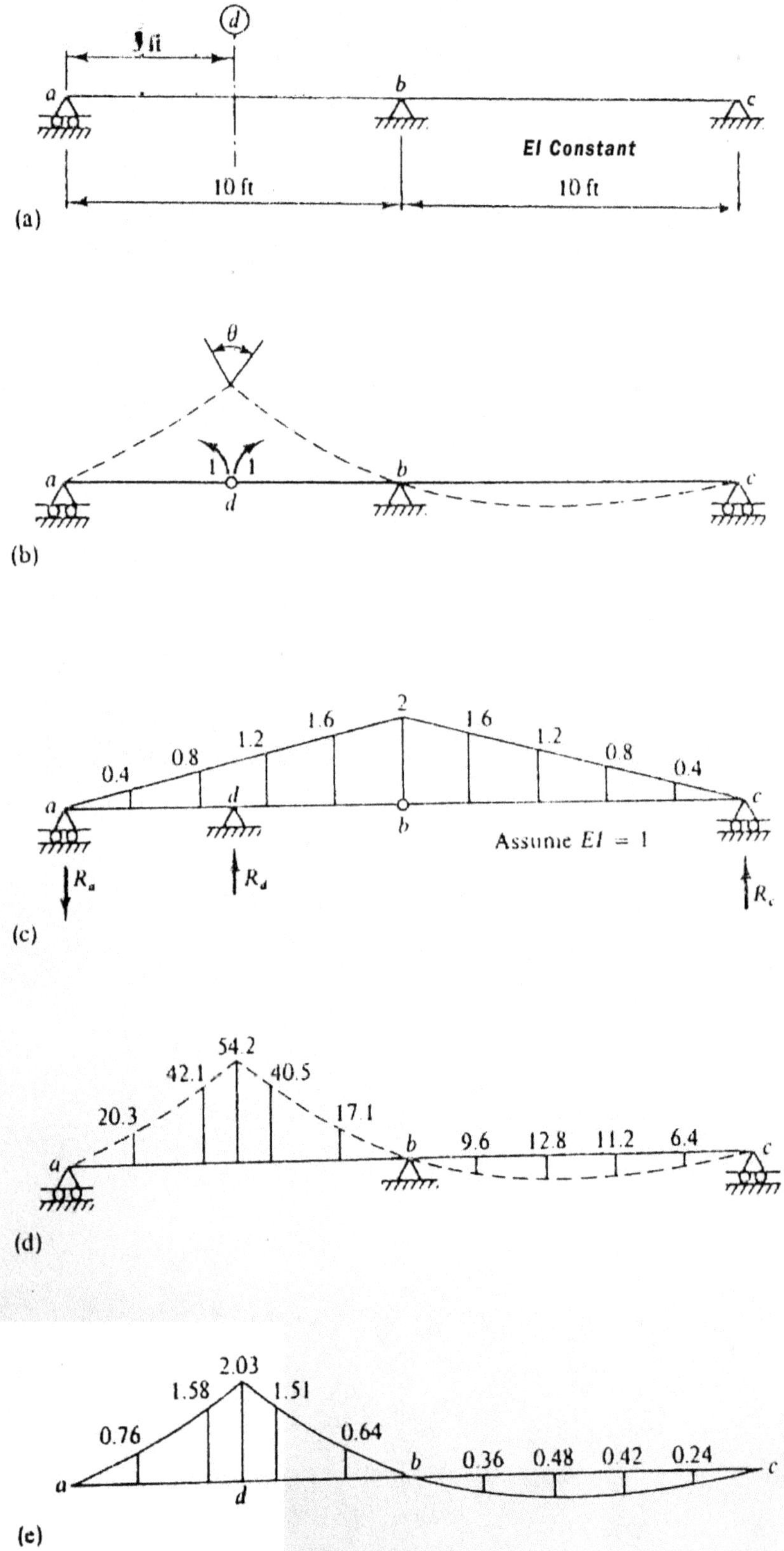

Figure 10-20

Problem 2-b

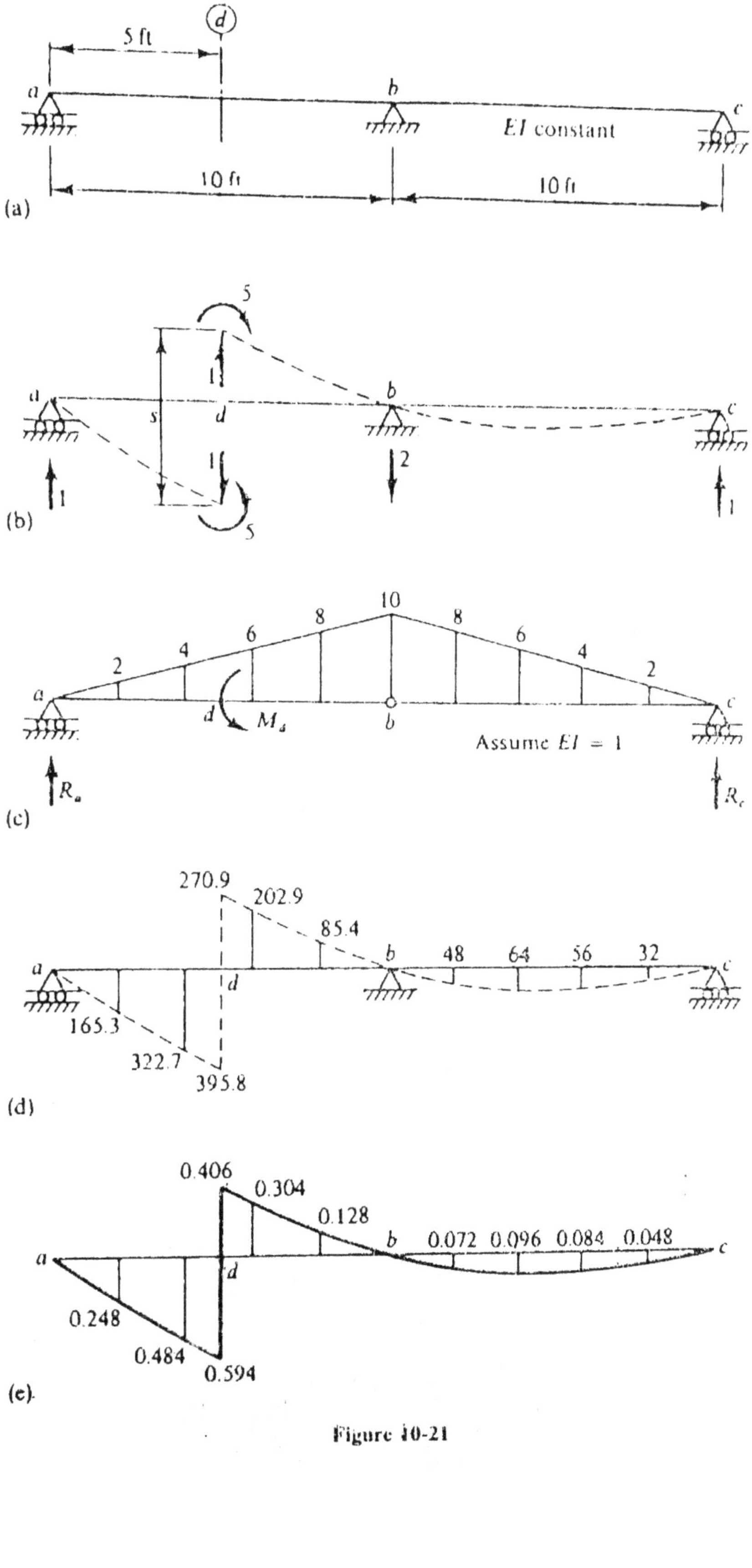

Figure 10-21

(3) Compute the ordinates at 2-m intervals of the influence line for the reaction at A of the beam shown in the figure. [H3]

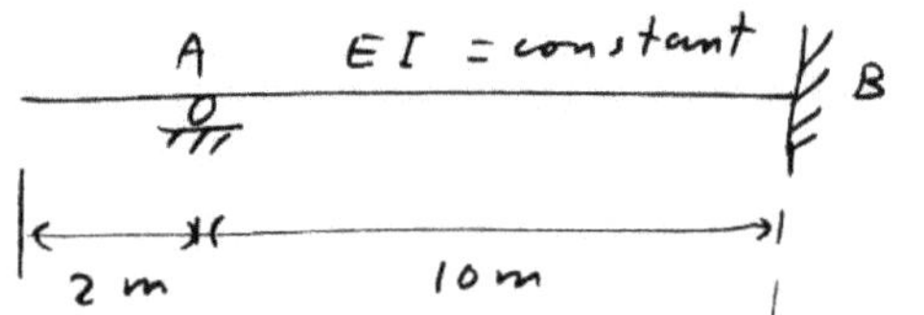

Solution:

$$R_A = -\frac{\delta x_{A0}}{f_{AA}}$$

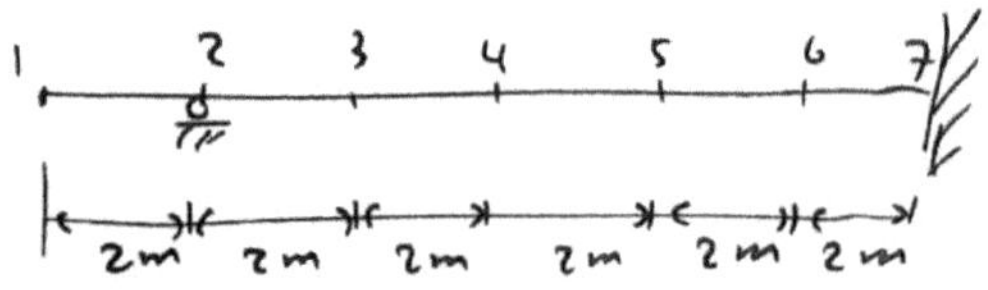

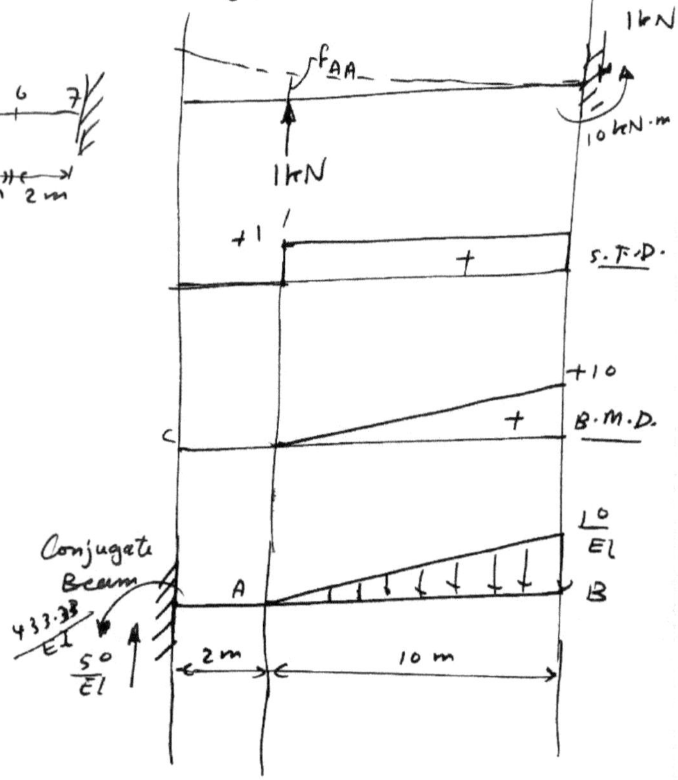

At point 1:

$$\delta x_{A0} = -\frac{433.33}{EI}.$$

At point 2:

$$\delta x_{A0} = -\frac{433.33}{EI} + \frac{50}{EI}(2)$$

$$= -\frac{333.33}{EI}.$$

At point 3:

$$\delta x_{A0} = \frac{-433.33}{EI} + \frac{50}{EI}(4) - \frac{1}{2}(2)\left(\frac{2}{EI}\right)\left(\frac{2}{3}\right)$$

$$= -\frac{234.67}{EI}$$

At point 4:

$$\delta x_{A0} = \frac{-433.33}{EI} + \frac{50}{EI}(6) - \frac{1}{2}(4)\left(\frac{4}{EI}\right)\left(\frac{4}{3}\right) = -\frac{144.0}{EI}.$$

At point 5:

$$\delta x_{A0} = -\frac{433.33}{EI} + \frac{50}{EI}(8) - \frac{1}{2}(6)\left(\frac{6}{EI}\right)\left(\frac{6}{3}\right) = -\frac{69.33}{EI}.$$

At point 6:

$$\delta x_{A0} = \frac{-433.33}{EI} + \frac{50}{EI}(10) - \frac{1}{2}(8)\left(\frac{8}{EI}\right)\left(\frac{8}{3}\right) = -\frac{18.86}{EI}.$$

H4

At point 7 :

$$\delta_{x_{A_0}} = \frac{-433.33}{EI} + \frac{50}{EI}(12) - \frac{1}{2}(10)\left(\frac{10}{EI}\right)\left(\frac{10}{3}\right) = 0 .$$

Note : $f_{AA} = -\delta_{x_{A_0}}$ at point 2 (A)

$$= +\frac{333.3}{EI} .$$

Influence line ordinate on $R_A = -\dfrac{\delta_{x_{A_0}}}{f_{AA}}$

point	ordinate
1	1.300
2	1.000
3	0.704
4	0.432
5	0.208
6	0.056
7	0.0

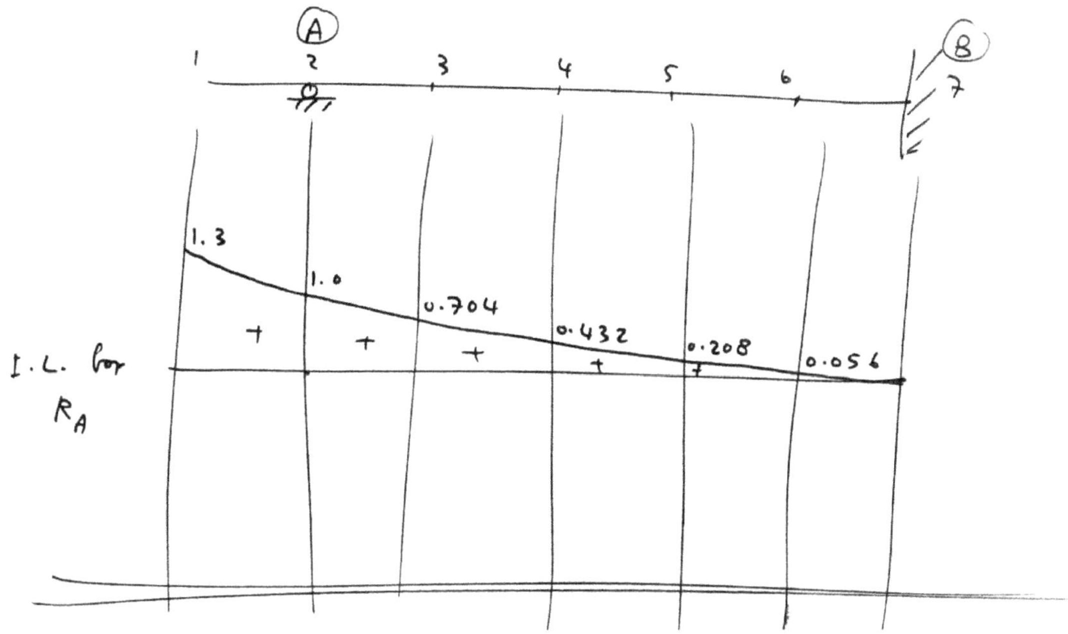

(4) Draw qualitative influence lines for : (HS

(a) the reaction at A and B.

(b) the shear just to the left of B.

(c) the shear just to the right of B.

(d) the moment at B.

(e) the shear and moment at a section in the middle of span DE.

Solution : See attached solution.

extremely useful in determining the loading patterns for design. Figure 10-17 shows typical influence lines for a five-span continuous beam.

The sketches in Fig. 10-17 indicate that if a maximum R_a is desired then spans ab, cd, and ef should be loaded; if the maximum values of reaction, of shear, and of bending moment at b are desired, then spans ab, bc, and de should be loaded.

Problem (4)

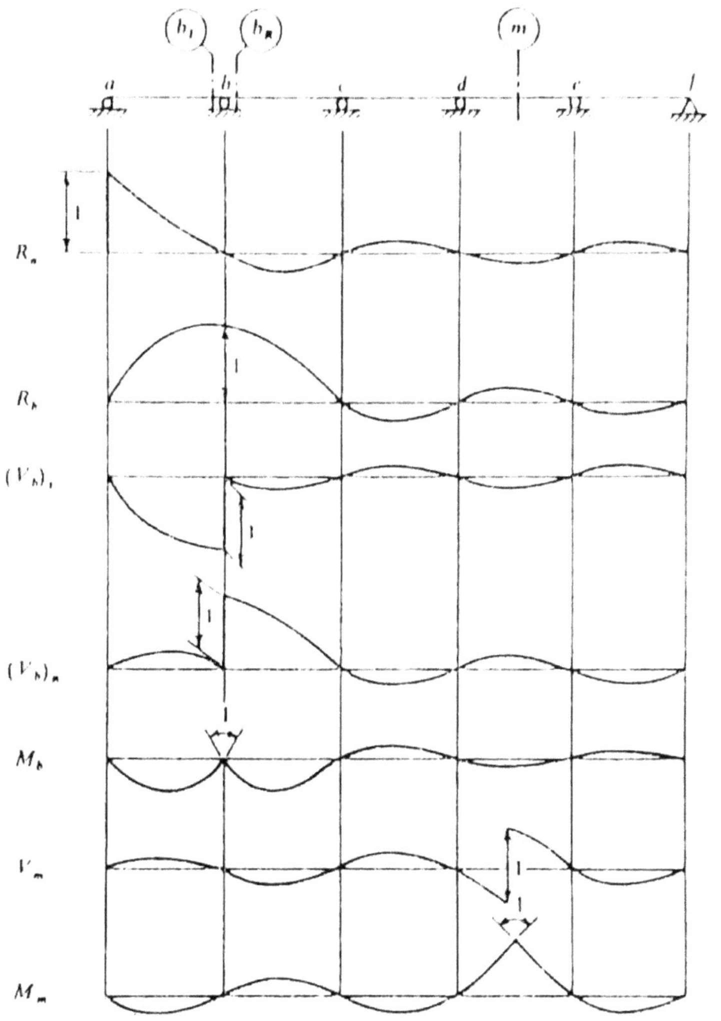

Figure 10-17

Figure 10-18(a) shows the influence line for the positive moment at the midspan section of $A3$–$B3$ of the frame shown. The uniform loading pattern for obtaining the maximum positive moment of this section is shown in Fig. 10-18(b).

Numerical examples for influence lines of statically indeterminate structures follow.

Structural Analysis (2)

Exam -2- / Solutions

① Using the slope-deflection method, write the shear equation (in terms of the joint moments) for the frame shown in the figure. Do not solve the problem.

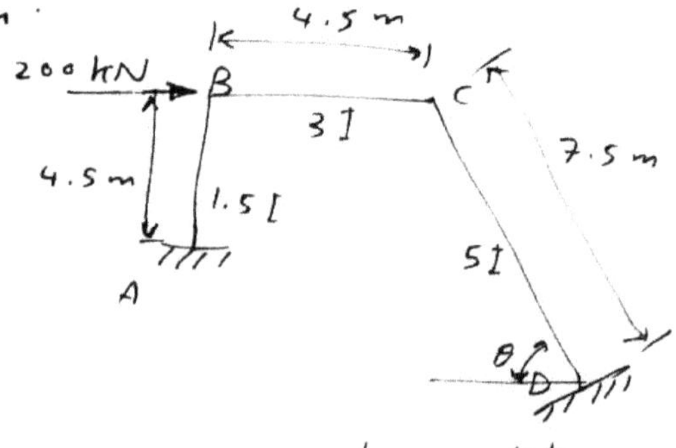

$\tan\theta = 4/3$.

Solution:

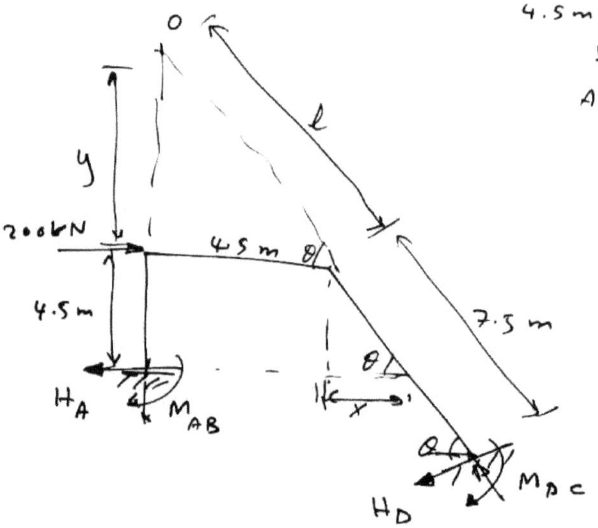

$$\frac{4}{3} = \tan\theta = \frac{4.5}{x} \Rightarrow x = 3.375 \text{ m}.$$

$$\frac{4}{3} = \tan\theta = \frac{y}{4.5} \Rightarrow y = 6 \text{ m}.$$

$$\therefore \; l = \sqrt{(4.5)^2 + (6)^2} = 7.5 \text{ m}.$$

$+\circlearrowleft \sum M_o = 0$:

$$-M_{AB} - M_{DC} + 200\,(6) - H_A\,(10.5) - H_D\,(15) = 0$$

but

$$H_A = \frac{-M_{AB} - M_{BA}}{4.5}$$

$$H_D = \frac{-M_{CD} - M_{DC}}{7.5}$$

$$\Rightarrow -M_{AB} - M_{DC} + 1200 + 15\left(\frac{M_{CD} + M_{DC}}{7.5}\right) + 10.5\left(\frac{M_{AB} + M_{BA}}{4.5}\right) = 0$$

$$\boxed{1.333\,M_{AB} + 2.333\,M_{BA} + M_{DC} + 2\,M_{CD} + 1200 = 0}$$

② Draw the bending moment diagram for the frame (2 shown in the figure.

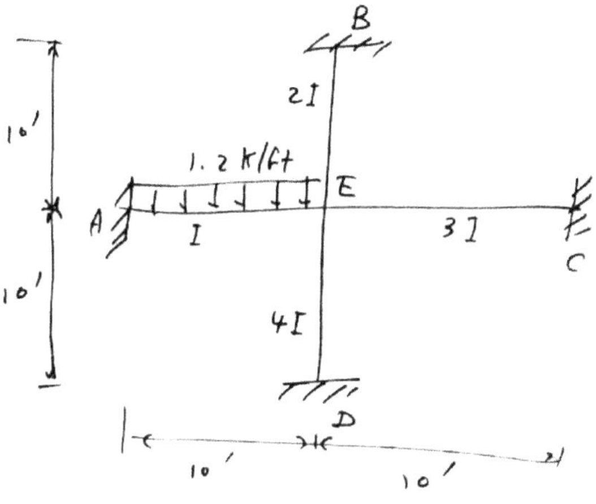

Solution :

Joint E :

$K_{EA} = I/10$

$K_{EB} = 2I/10$

$K_{EC} = 3I/10$

$K_{ED} = 4I/10$

$\Sigma K_E = \frac{I}{10} + \frac{2I}{10} + \frac{3I}{10} + \frac{4I}{10} = I.$

$DF_{EA} = \frac{I/10}{I} = 0.1$

$DF_{EB} = \frac{2I/10}{I} = 0.2$

$DF_{EC} = 0.3$

$DF_{ED} = 0.4$

$M^F_{AE} = M^F_{EA} = \frac{wl^2}{12}$

$= \frac{(1.2)(10)^2}{12}$

$= 10$ k-ft.

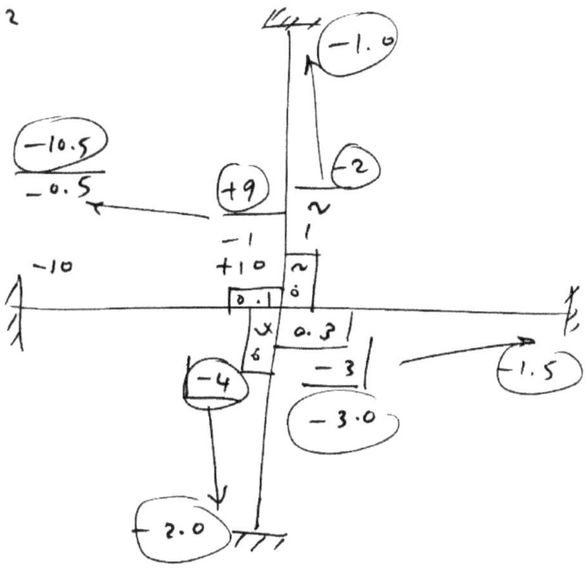

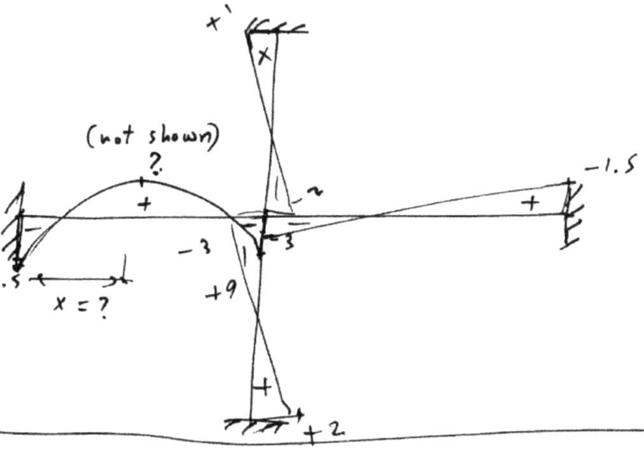

B.M.D.
(k-ft)

3

(3) Find the equation $y = f(x)$ of the influence line for the reaction at support B of the continuous beam shown in the figure. Do not calculate the ordinates or plot the curve.

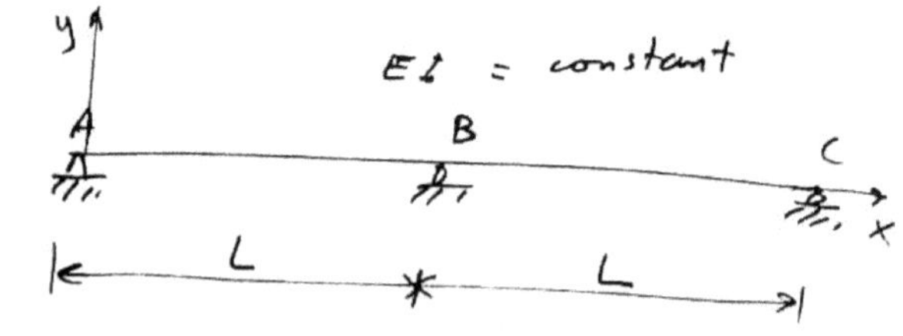

Solution :

Remove the support at B and place a unit load along the line of the reaction.

$$0 = \delta_{BO(x)} + R_B f_{BB}$$

$$0 = \delta_{xBO} + R_B f_{BB}$$

$$\Rightarrow \boxed{R_B = -\frac{\delta_{xBO}}{f_{BB}}}$$

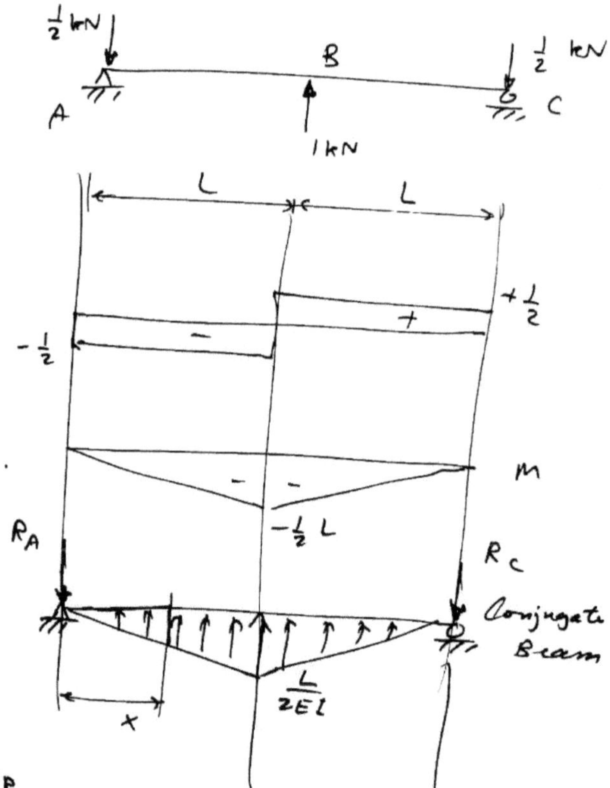

For the conjugate beam:

$$R_A = \frac{1}{2} L \cdot \frac{L}{2EI} = \frac{L^2}{4EI}$$

$$R_C = \frac{L^2}{4EI} \quad \text{(by symmetry)}.$$

$$\delta_{xBO} = \frac{1}{2} x \cdot \frac{x}{2EI}\left(\frac{1}{3}x\right) - R_A\, x$$

$$= \frac{x^3}{12EI} - \frac{L^2 x}{4EI}$$

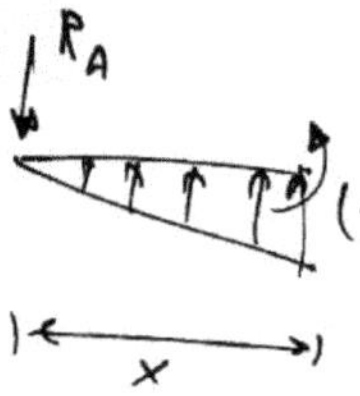

$$\left(\frac{x}{L}\right)\left(\frac{L}{2EI}\right) = \frac{x}{2EI}$$

$$f_{BB} = \delta_{xBO} \quad (\text{at } x = L)$$

$$= \frac{L^3}{12EI} - \frac{L^3}{4EI}$$

$$= \left| -\frac{L^3}{6EI} \right| = +\frac{L^3}{6EI}$$

$$\Rightarrow R_B = -\left(+\frac{6EI}{L^3}\right)\left(\frac{x^3}{12EI} - \frac{L^2 x}{4EI}\right) = -\frac{6EI}{L^3}\left(\frac{x}{12EI}\right)(x^2 - 3L^2) = \frac{x(x^2-3L^2)}{2L^3}$$

$$\Rightarrow \boxed{y = \frac{-x^3 + 3L^2 x}{2L^3}} = \frac{3L^2 x - x^3}{2L^3}, \quad 0 \le x \le L.$$

Structural Analysis (2)
Final Exam / Solutions

(1)

(25%)

① Using the method of consistent deformations, determine the fixed-end moments for the beam shown in the figure. Neglect axial forces. Take EI = constant.

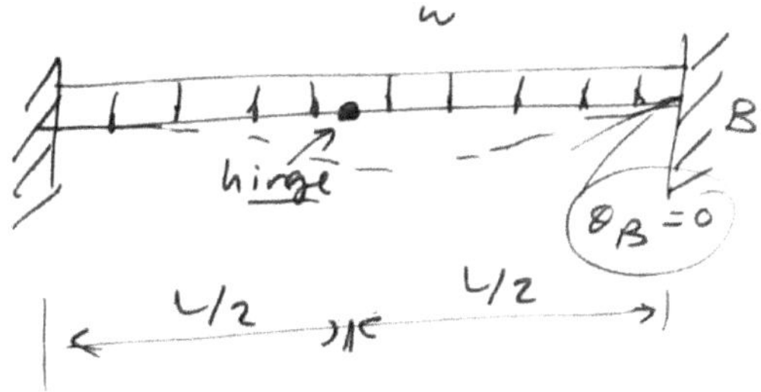

Solution:

Choose M_B as the redundant.

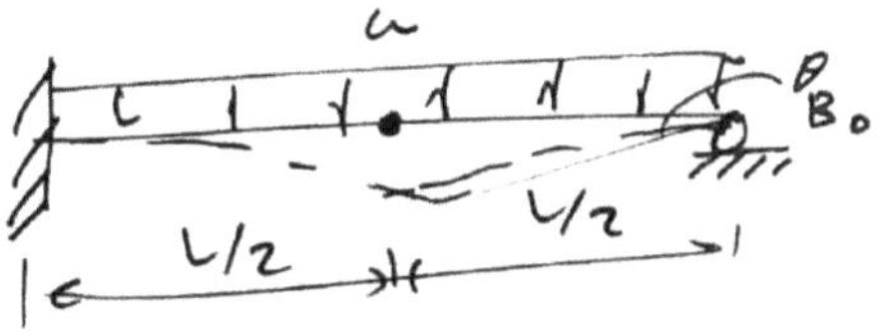

$$\theta_B = \theta_{B_0} + M_B f_{BB} = 0$$

+

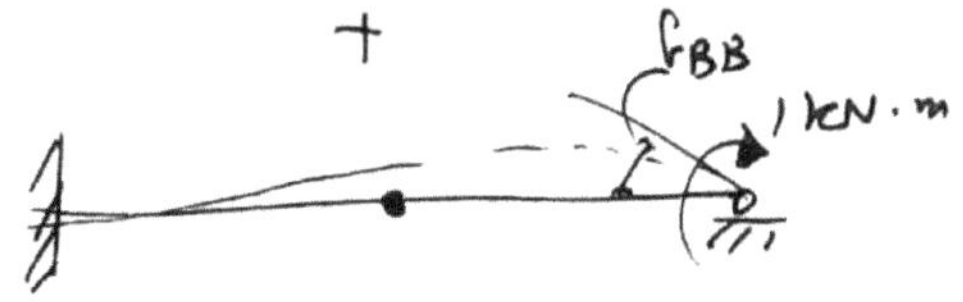

Step 1: Find θ_{B_0} from the primary structure:

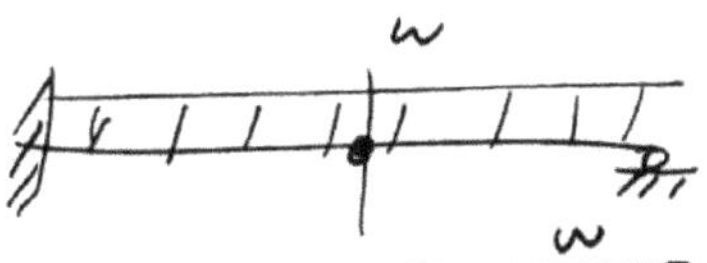

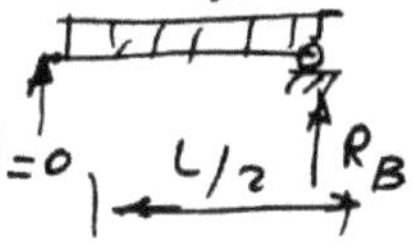

$\circlearrowleft \sum M_c = 0$:

$R_B\left(\frac{L}{2}\right) - w\left(\frac{L}{2}\right)\left(\frac{L}{4}\right) = 0$

$R_B = \frac{2}{L} \cdot \frac{wL^2}{8}$

$R_B = \frac{wL}{4}$ ✓

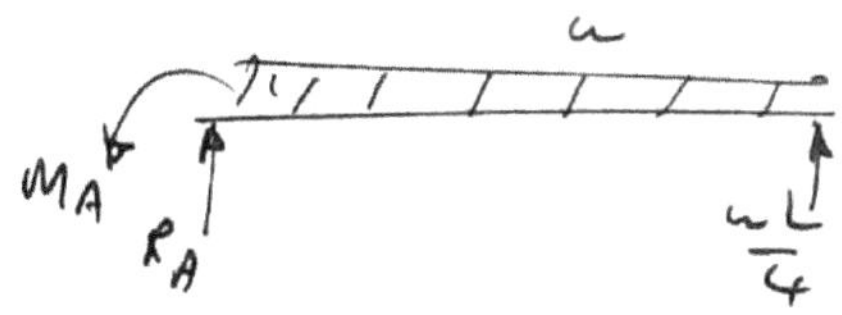

$R_A = wL - \frac{wL}{4} = \frac{3wL}{4}$ ✓

$M_A = wL\left(\frac{L}{2}\right) - \frac{wL}{4}(L)$

$= \frac{1}{2}wL^2 - \frac{1}{4}wL^2 = \frac{1}{4}wL^2$ ✓

12

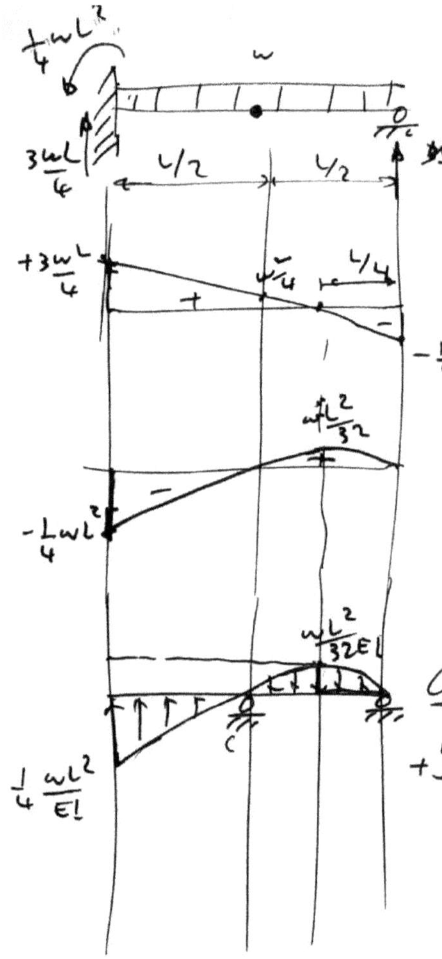

$$\frac{x}{\frac{1}{4}} = \frac{L-x}{3/4}$$

$$3x = L - x$$
$$4x = L \Rightarrow x = L/4.$$

$$M_0 = -\frac{1}{4}wL^2 + \frac{1}{2}\left(\frac{3wL}{4}\right)\left(\frac{3L}{4}\right)$$
$$= -\frac{1}{4}wL^2 + \frac{9wL^2}{32}$$
$$= \frac{wL^2}{32}$$

$$M_c = -\frac{1}{4}wL^2 + \frac{1}{2}\left(\frac{3wL}{4} + \frac{wL}{4}\right)\left(\frac{L}{2}\right)$$
$$= -\frac{1}{4}wL^2 + \frac{1}{4}wL^2 = 0.$$

$$\frac{3}{4}\left(\frac{3L}{4}\right) - \frac{L}{4} = \frac{9L}{16} - \frac{L}{4} = \frac{5L}{16}$$

$+\circlearrowleft \Sigma M_c = 0:$

$$\theta_{B0} = R_B = 2\left(\frac{2}{3}\right)\left(\frac{L}{4}\right)\left(\frac{wL^2}{32EI}\right)\left(\frac{L}{4}\right) + \left[\frac{1}{3}\left(\frac{9wL^2}{32EI}\right)\left(\frac{3}{4}L\right)\left(\frac{5L}{16}\right) + \frac{1}{3}\left(\frac{wL^2}{32EI}\right)\left(\frac{L}{4}\right)\left(\frac{1}{4}\cdot\frac{L}{4}\right) - \frac{wL^2}{32EI}\left(\frac{L}{2}\right)\left(\frac{L}{4}\right)\right] \Big/ L/2$$

$$\Rightarrow \theta_{B0} = \left(\frac{wL^3}{3(32EI)}\cdot\frac{L}{4} + \frac{9wL^3}{4(32EI)}\cdot\frac{5L}{16} + \frac{wL^3}{12(32EI)}\cdot\frac{L}{16} - \frac{wL^3}{2(32EI)}\cdot\frac{L}{4}\right)\left(\frac{2}{L}\right)$$

$$= \frac{2wL^3}{32EI}\left[\frac{1}{4}\cdot\frac{1}{3} + \frac{9}{4}\left(\frac{5}{16}\right) + \frac{1}{12}\left(\frac{1}{16}\right) - \frac{1}{2}\left(\frac{1}{4}\right)\right]$$

$$= \frac{wL^3}{16EI}\left(\frac{1}{12} + \frac{45}{4(16)} + \frac{1}{12(16)} - \frac{1}{8}\right)$$

$$= \frac{wL^3}{16EI}\left(\frac{16 + 3(45) + 1 - 24}{12(16)}\right)$$

$$= \frac{wL^3}{24EI}.$$

3

Step 2: Find f_{BB} from the second structure:

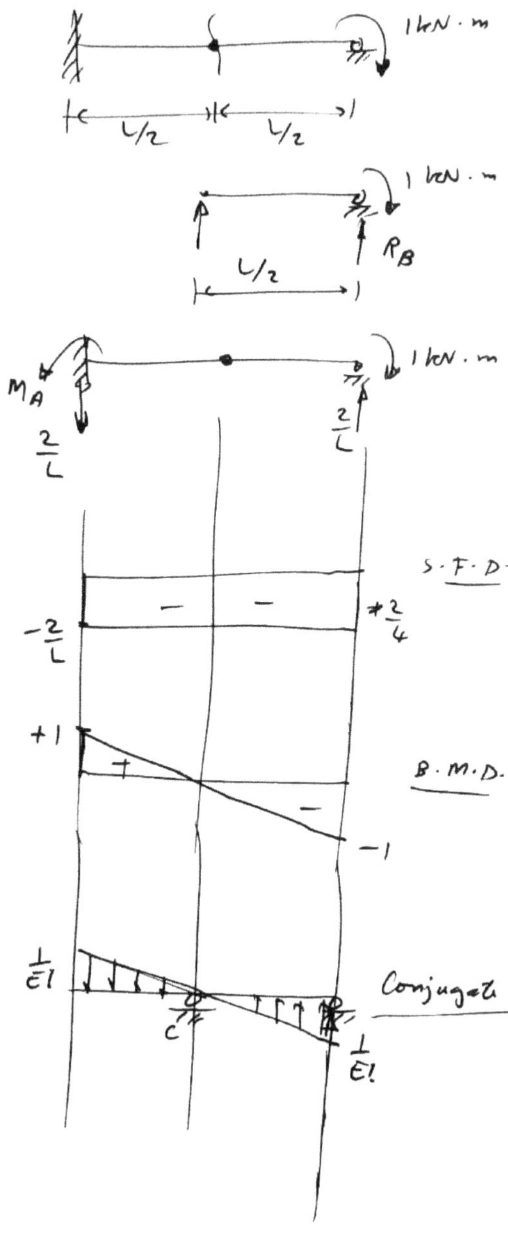

$+\circlearrowleft \Sigma M_c = 0$:

$R_B\left(\frac{L}{2}\right) - 1 = 0$

$R_B = \frac{2}{L}$ ✓

$M_A = 1 - \frac{2}{L}(L) = -1$ kN·m.

Conjugate Beam

For the conjugate beam:

$\circlearrowleft \Sigma M_c = 0$:

$R_B\left(\frac{L}{2}\right) + \frac{1}{2}\left(\frac{1}{EI}\right)\left(\frac{L}{2}\right)\left(\frac{2}{3}\cdot\frac{L}{2}\right) + \frac{1}{2}\left(\frac{1}{EI}\right)\left(\frac{L}{2}\right)\left(\frac{2}{3}\cdot\frac{L}{2}\right) = 0$

$R_B\left(\frac{L}{2}\right) + \frac{1}{EI}\cdot\frac{L^2}{6} = 0$

$R_B = -\frac{L^2}{6EI}\cdot\frac{2}{L}$

$f_{BB} \Rightarrow \boxed{R_B = -\frac{L}{3EI}}$ ✓

Step 3: Equation of Consistent Deformations:

$\theta_{B0} + M_B f_{BB} = 0$

$\Rightarrow -\frac{wL^3}{24EI} + M_B\left(\frac{L}{3EI}\right) = 0 \Rightarrow M_B = \frac{wL^3}{24EI}\cdot\frac{3EI}{L}$

$\Rightarrow \boxed{M_B = \frac{wL^2}{8}}$ = $\frac{wL^2}{8}$.

(2) (20%) Draw the bending moment diagram for the two-hinged semi-circular arch with radius (15 m) shown in the figure. Use the method of least work (Castigliano's Second Theorem): Take EI = constant. [4

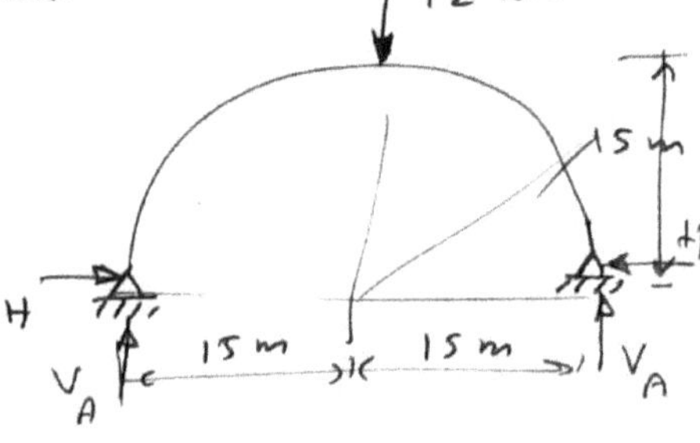

Solution:

Choose H as the redundant.

$V_A = \frac{12}{2} = 6$ kN (by symmetry).

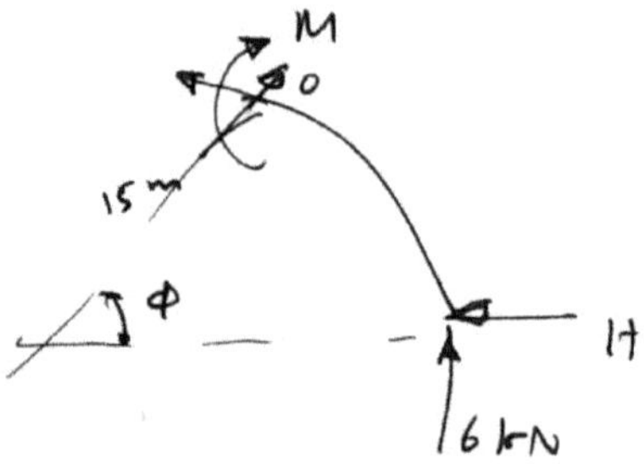

$$+\circlearrowleft \Sigma M_o = 0: \quad -M - H(15\sin\phi) + 6(15 - 15\cos\phi) = 0$$

$$M = 6(15)(1-\cos\theta) - 15H\sin\theta$$

$$M = 90(1-\cos\theta) - 15H\sin\theta$$

$$\frac{\partial M}{\partial H} = -15\sin\theta$$

$$ds = R\,d\phi = 15\,d\phi$$

$$\frac{\partial U}{\partial H} = 0 = \int \frac{M}{EI}\frac{\partial M}{\partial H}\,ds = 0$$

$$\Rightarrow \int M\frac{\partial M}{\partial H} \cdot 15\,d\phi = 0$$

$$(2) \cdot 15\int_0^{\pi/2} \left(90(1-\cos\theta) - 15H\sin\phi\right)(-15\sin\phi)\,d\phi = 0$$

$$\int_0^{\pi/2} \left[90(1-\cos\theta) - 15H\sin\phi\right]\sin\phi\,d\phi = 0$$

$$\int_0^{\pi/2} \left(90\sin\phi - 90\sin\phi\cos\phi - 15H\sin^2\phi\right)d\phi = 0$$

$$\int_0^{\pi/2} \left[90\sin\phi - \frac{90}{2}\sin 2\phi - \frac{15H}{2}(1-\cos 2\phi)\right] d\phi = 0$$

$\sin\phi\cos\phi = \frac{\sin 2\phi}{2}$

$\cos 2\phi = \cos^2\phi - \sin^2\phi$

$= 1 - \sin^2\phi - \sin^2\phi$

$\cos 2\phi = 1 - 2\sin^2\phi$

$\sin^2\phi = \frac{1-\cos 2\phi}{2}$

5

$$\left[-90\cos\phi + \frac{45}{2}\cos 2\phi - \frac{15H}{2}\left(\phi - \frac{1}{2}\sin 2\phi\right)\right]_0^{\pi/2} = 0$$

$$-90(0) + \frac{45}{2}(-1) - \frac{15H}{2}\left(\frac{\pi}{2} - 0\right) + 90 - \frac{45}{2}(1) + \frac{15H}{2}(0-0) = 0$$

$$-45 - 15\frac{H}{2}\cdot\frac{\pi}{2} + 90 = 0$$

$$\frac{15H\pi}{4} = 45 \quad \Longrightarrow \quad \boxed{H = \frac{12}{\pi}}$$

③ (15%) Using the slope-deflection method, write the shear equation for the frame shown in the figure in terms of the joint moment. Do NOT try to solve the problem. 16

Solution :

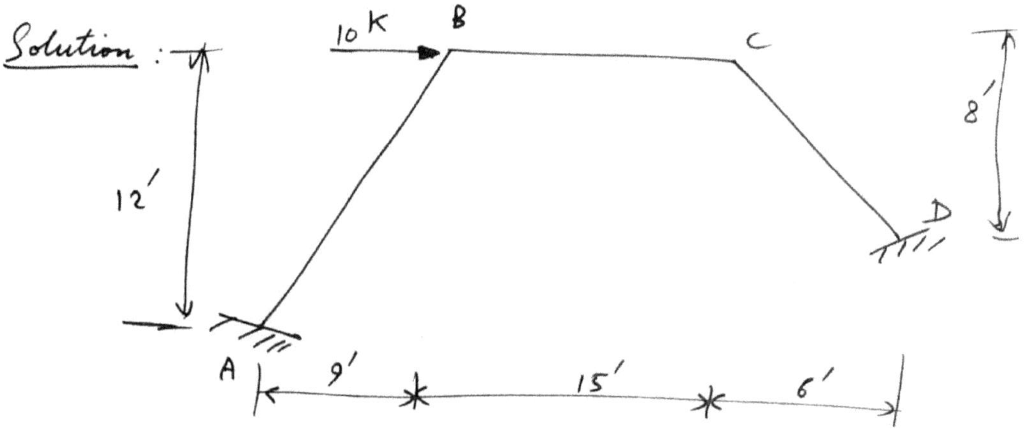

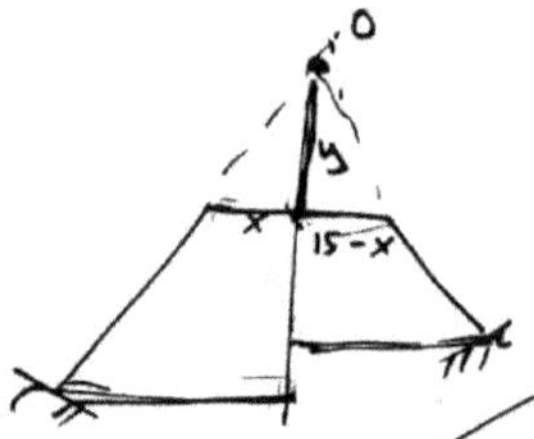

$$\frac{x}{y} = \frac{x+9}{y+12} \longrightarrow (1)$$

$$\frac{15-x}{y} = \frac{15-x+6}{y+8} \longrightarrow (2)$$

Solve the two equations simultaneously for x and y :

$$x(y+12) = y(x+9)$$

$$\Rightarrow \quad \cancel{xy} + 12x = \cancel{xy} + 9y \Longrightarrow \boxed{12x = 9y}$$

$$(15-x)(y+8) = (21-x)y$$

$$15y + 120 - \cancel{xy} - 8x = 21y - \cancel{xy}$$

$$\boxed{120 - 8x = 6y}$$

7

$12x = 9y$

$x = \frac{9y}{12}$

$8x = 8\left(\frac{9}{12}y\right) = 6y \Rightarrow 120 - 6y = 6y$

$120 = 12y \Rightarrow \boxed{y = 10'}$

$\Rightarrow x = \frac{9}{12}(10) = \boxed{7.5 \text{ ft}}$

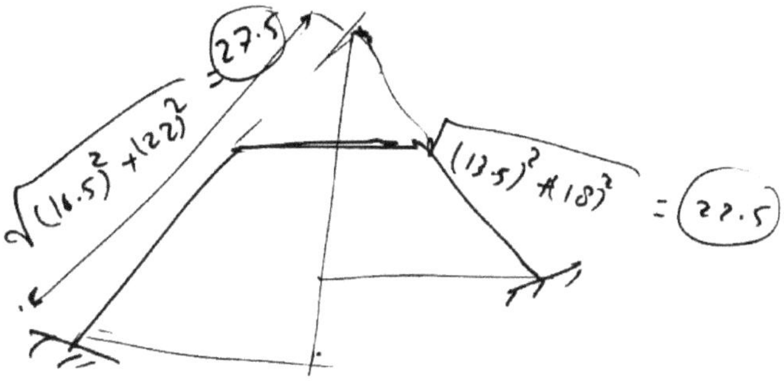

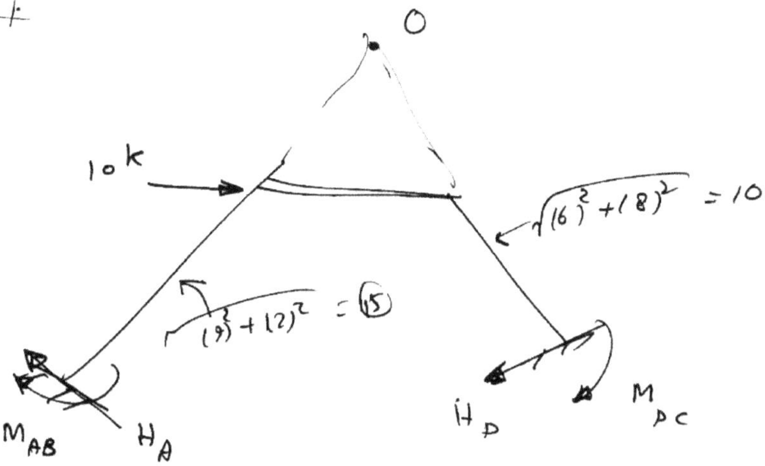

Shear Equation:

$+\circlearrowleft\ \Sigma M_o = 0:$

$10(10) - H_A(27.5) - H_D(27.5) - M_{AB} - M_{DC} = 0$

but $H_A = \dfrac{-M_{AB} - M_{BA}}{15}$

$H_D = \dfrac{-M_{CD} - M_{DC}}{10}$

$100 + 27.5\left(\dfrac{M_{AB} + M_{BA}}{15}\right) + 22.5\left(\dfrac{M_{CD} + M_{DC}}{10}\right) - M_{AB} - M_{DC} = 0$

$\boxed{0.833 M_{AB} + 1.833 M_{BA} + 2.25 M_{CD} + 1.25 M_{DC} + 100 = 0}$

(4) (15%) Draw the bending moment diagram for the frame shown in the figure. Use the moment distribution method: Take EI = constant. [8]

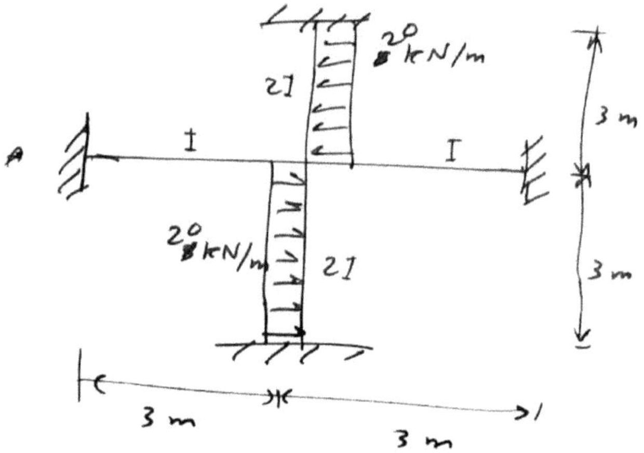

Solution:

$$\Sigma K = I/3 + I/3 + 2I/3 + 2I/3$$

$$= 6I/3 = 2I.$$

$$DF_1 = \frac{I/3}{2I} = \frac{1}{6}.$$

$$DF_2 = \frac{2I/3}{2I} = \frac{2}{6} = \frac{1}{3}.$$

$$\frac{wL^2}{12} = \frac{20(3)^2}{12} = 15 \text{ kN·m}$$

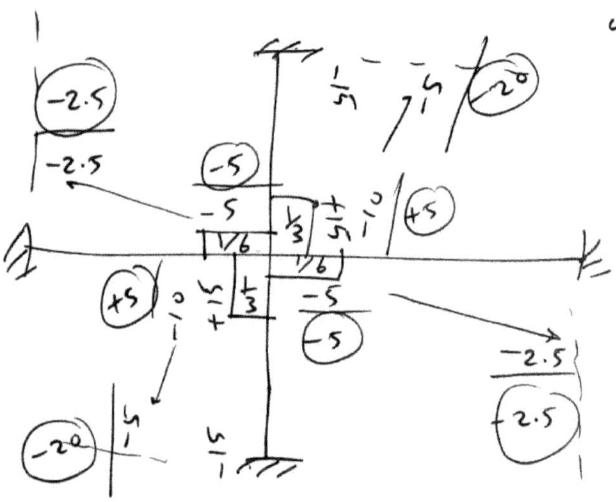

B.M.D.

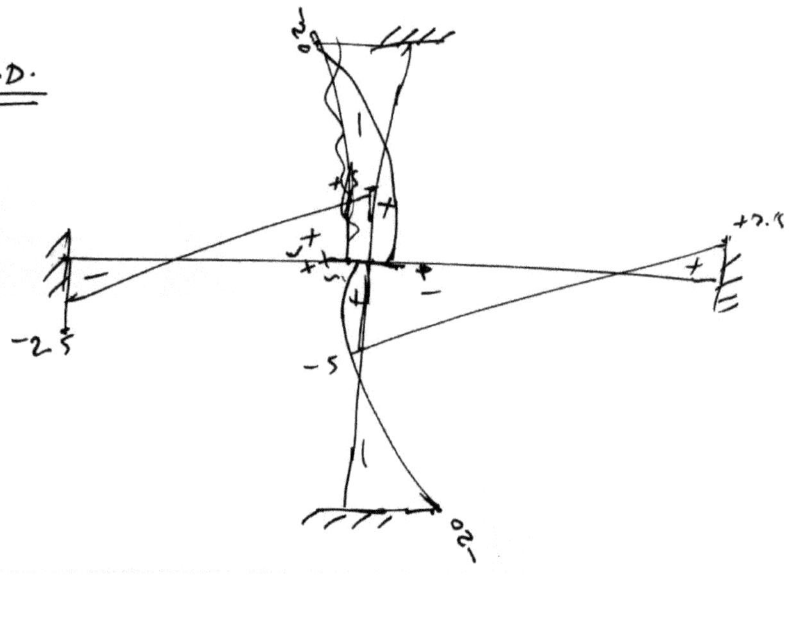

(5) (25%) Draw the influence line for the reaction at A showing numerical ordinates at 1 m intervals for the continuous beam shown in the figure. L9

Solution:

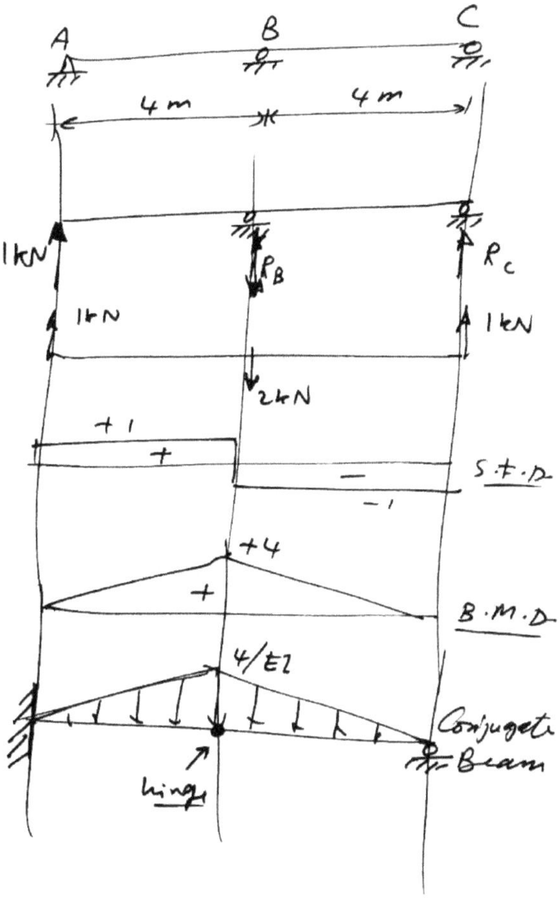

Choose R_A as the redundant.

$+\circlearrowleft \Sigma M_B = 0:$

$R_C(4) - 1(4) = 0$

$R_C = 1 \text{ kN} \uparrow.$

$+\uparrow \Sigma F_y = 0:$

$1 - R_B + 1 = 0$

$R_B = 2 \text{ kN} \downarrow.$

For the conjugate beam:

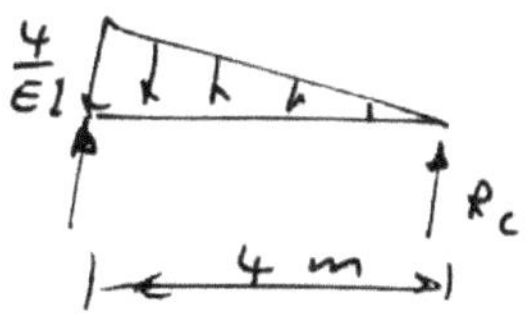

$$R_c(4) - \frac{1}{2}\left(\frac{4}{EI}\right)(4)\left(\frac{4}{3}\right) = 0$$

$$R_c = \frac{8}{3EI} \uparrow.$$

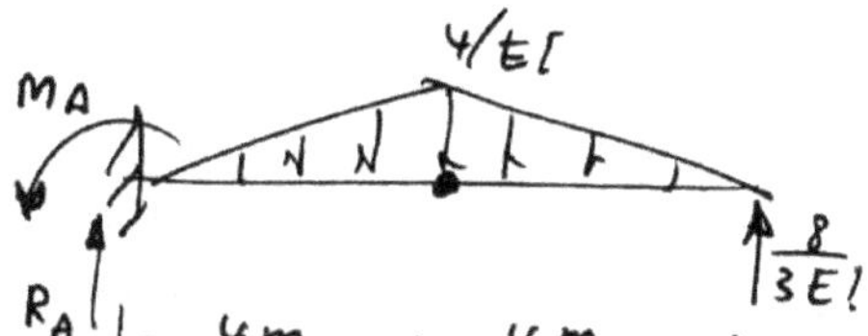

$+\uparrow \Sigma F_y = 0:$

$$R_A + \frac{8}{3EI} - \frac{1}{2}\left(\frac{4}{EI}\right)(8) = 0$$

$$R_A = \frac{16}{EI} - \frac{8}{3EI} = \frac{48 - 8}{3EI} = \frac{40}{3EI}.$$

$$M_A = \frac{1}{2}\left(\frac{4}{EI}\right)(8)(4) - \frac{8}{3EI}(8) = \frac{64}{EI} - \frac{64}{3EI} = \frac{128}{3EI}$$

10

For $0 \le x \le 4$: (right)

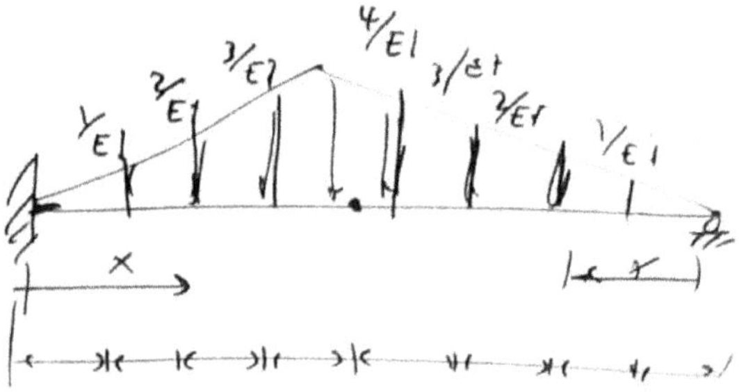

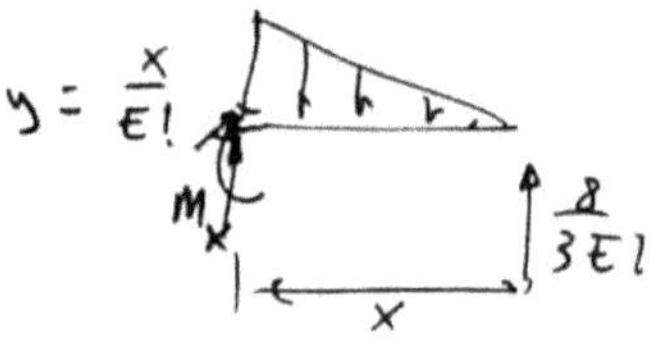

$$\delta_{x_{A_0}} = \frac{8}{3EI}x - \frac{1}{2} \times \left(\frac{x}{EI}\right)\left(\frac{x}{3}\right) = \boxed{\frac{8x}{3EI} - \frac{x^3}{6EI}}$$

For $0 \le x \le 4$: (left)

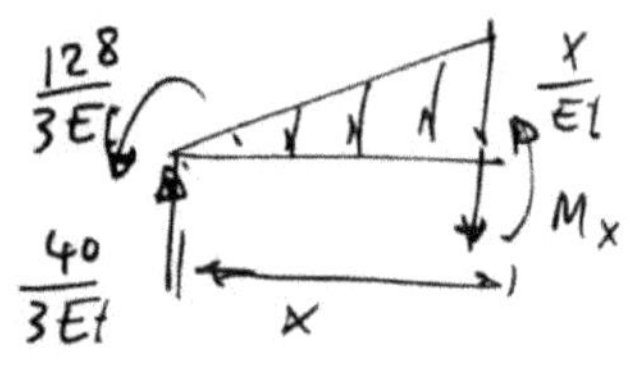

$$\delta_{x_{A_0}} = \frac{40}{3EI}x - \frac{128}{3EI} - \frac{1}{2}\left(\frac{x}{EI}\right) \times \left(\frac{x}{3}\right)$$

$$= \boxed{\frac{40x}{3EI} - \frac{128}{3EI} - \frac{x^3}{6EI}}$$

At $x = 0$: $\delta_{A_0} = f_{AA} = \dfrac{-128}{3EI}$.

$$\Rightarrow R_A = -\frac{\delta_{x_{A_0}}}{f_{AA}} = \frac{3EI}{128}\begin{cases} \dfrac{8x}{3EI} - \dfrac{x^3}{6EI} & \text{(right)} \\ \dfrac{40x}{3EI} - \dfrac{128}{3EI} - \dfrac{x^3}{6EI} & \text{(left)} \end{cases}$$

$$\boxed{R_A = \frac{-EI}{256}\begin{cases} 16x - x^3 & , x \text{ from right} \\ 80x - 256 - x^3 & , x \text{ from left} \end{cases}}$$

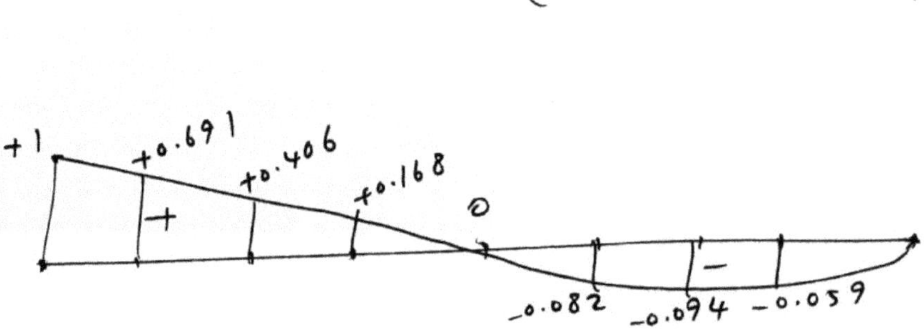

I.L. for R_A

Bonus Question: (+10%)

For each of the following structures, determine:

(a) degree of statical indeterminacy.

(b) degree of kinematical indeterminacy.

Solution:

① 

(a) 1 (b) 3 or 5.

② 

(a) 3 (b) 0

③ 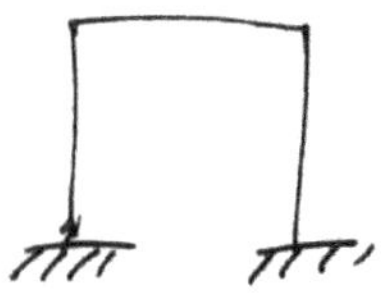

(a) 3 (b) 3 or 6

④ 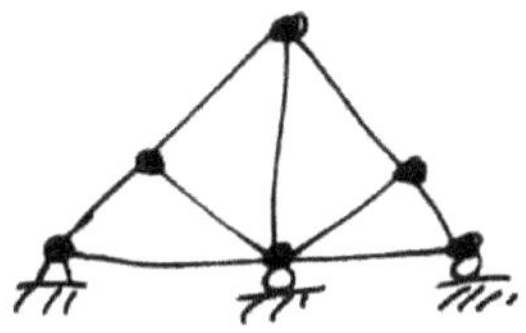

(a) 1 (b) 6 or 8.

(5) 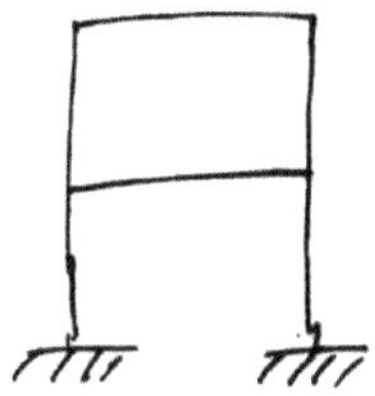

(a) 6 (b) 6 or 12

Fixed-End Moments

$\text{FEM}_{AB} = \frac{PL}{8}$

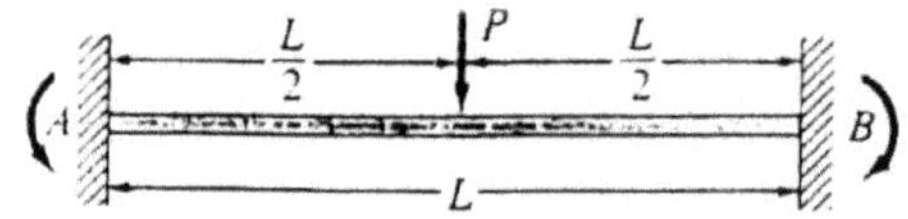

$\text{FEM}_{BA} = \frac{PL}{8}$

$\text{FEM}_{AB} = \frac{Pb^2a}{L^2}$

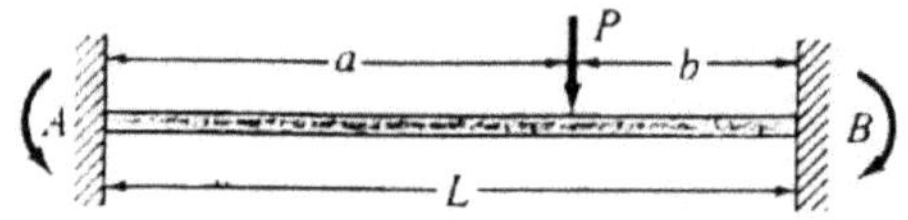

$\text{FEM}_{BA} = \frac{Pa^2b}{L^2}$

$\text{FEM}_{AB} = \frac{wL^2}{12}$

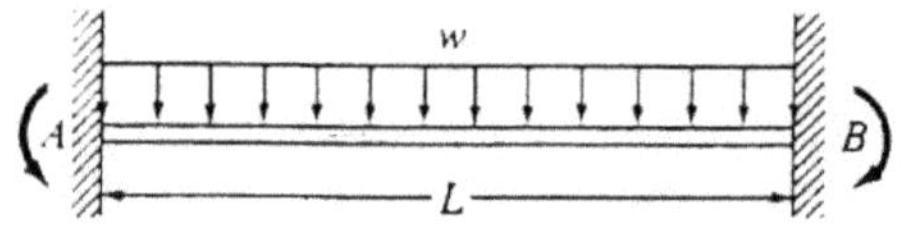

$\text{FEM}_{BA} = \frac{wL^2}{12}$

$\text{FEM}_{AB} = \frac{11\,wL^2}{192}$

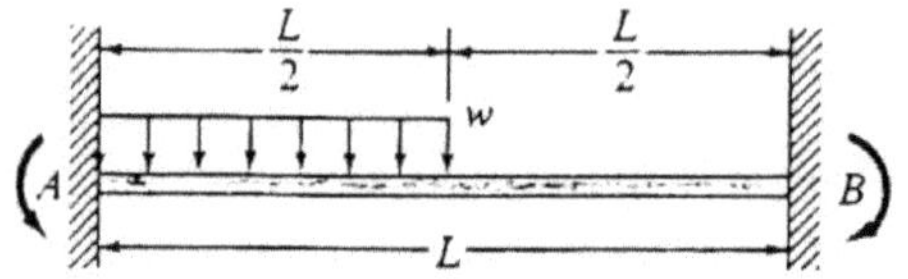

$\text{FEM}_{BA} = \frac{5wL^2}{192}$

$\text{FEM}_{AB} = \frac{wL^2}{20}$

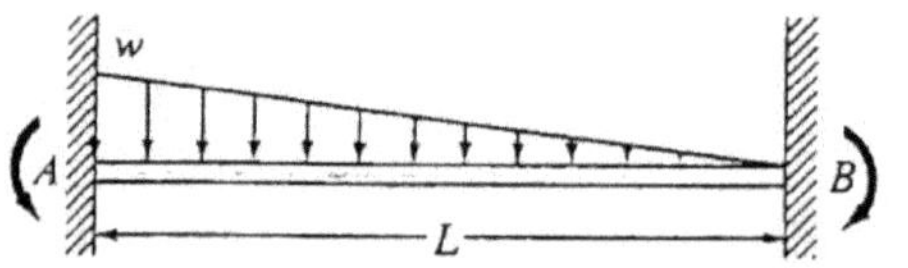

$\text{FEM}_{BA} = \frac{wL^2}{30}$

$\text{FEM}_{AB} = \frac{5wL^2}{96}$

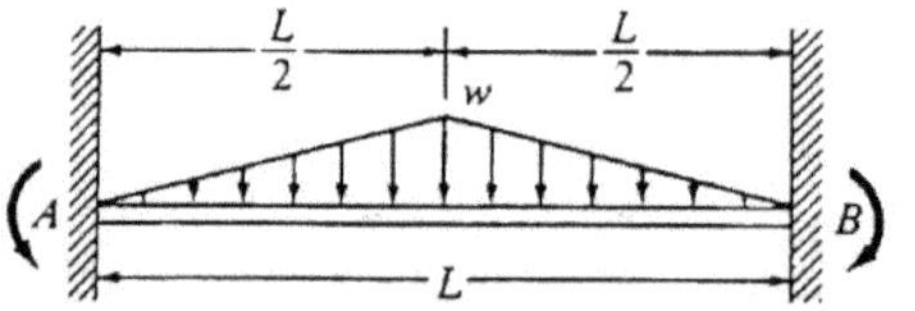

$\text{FEM}_{BA} = \frac{5wL^2}{96}$

$\text{FEM}_{AB} = \frac{6EI\Delta}{L^2}$

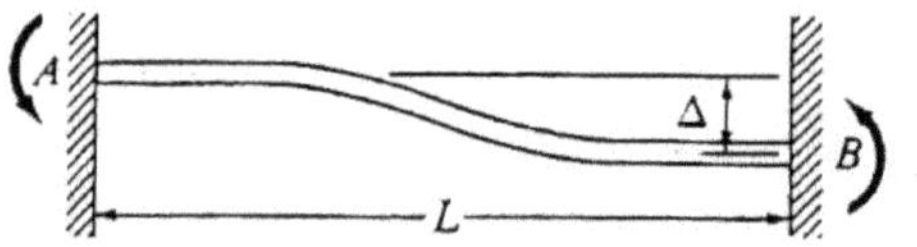

$\text{FEM}_{BA} = \frac{6EI\Delta}{L^2}$

$\text{FEM}_{AB} = M(1-k)(1-3k)$

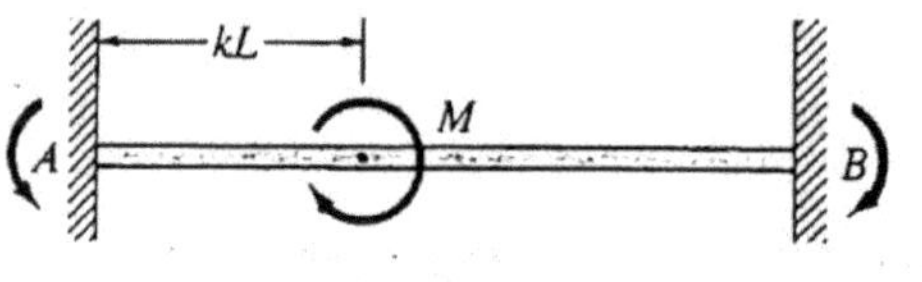

$\text{FEM}_{BA} = Mk(2-3k)$

$\text{FEM}_{AB} = \frac{wL^2}{60} \cdot k^2(10 - 10k + 3k^2)$

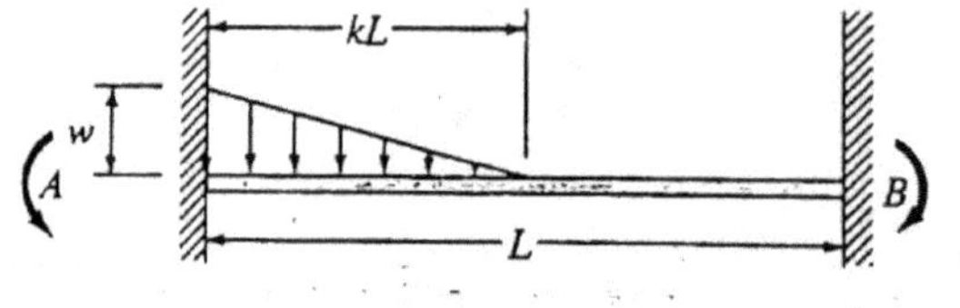

$\text{FEM}_{BA} = \frac{wL^2}{60} \cdot k^3(5-3k)$

www.ingramcontent.com/pod-product-compliance
Lightning Source LLC
LaVergne TN
LVHW070859210926
848482LV00016B/972